Orientations in Geochemistry

Panel on Orientations for Geochemistry
U.S. NATIONAL COMMITTEE FOR GEOCHEMISTRY
DIVISION OF EARTH SCIENCES
NATIONAL RESEARCH COUNCIL

NATIONAL ACADEMY OF SCIENCES
WASHINGTON, D.C.
1973

This report was supported by the National Science Foundation under Contract NSF C310, Task order 252.

Library of Congress Catalog Card Number 73-21348

ISBN 0-309-02147-2

Available from
Printing and Publishing Office
National Academy of Sciences
2101 Constitution Avenue
Washington, D.C. 20418

Printed in the United States of America

PANEL ON ORIENTATIONS FOR GEOCHEMISTRY

ROBERT M. GARRELS, University of Hawaii and Northwestern University
  Chairman
ROBERT A. DUCE, University of Rhode Island
EDWARD D. GOLDBERG, University of California at San Diego
STANLEY R. HART, Carnegie Institution of Washington
G. ROSS HEATH, Oregon State University
THOMAS C. HOERING, Carnegie Institution of Washington
HEINRICH D. HOLLAND, Harvard University
KEITH A. KVENVOLDEN, Ames Research Center, NASA
JOHN W. LARIMER, Arizona State University
FREDERICK T. MACKENZIE, Northwestern University
PHILIP M. ORVILLE, Yale University
LOUIS S. WALTER, Goddard Space Flight Center, NASA

Liaison
BEVAN M. FRENCH, National Science Foundation

Staff
WILLIAM L. PETRIE, *Executive Secretary*
JUNE R. GALKE, *Secretary*

U.S. NATIONAL COMMITTEE FOR GEOCHEMISTRY*

Contents

Preface   ix

Acknowledgments   xi

Summary   xiii

Introduction   1

Extraterrestrial Geochemistry   10

Solid Earth Geochemistry   31

Exogenic or Low-Temperature Aqueous Geochemistry   53

Organic Geochemistry   80

Geochemistry and Environmental Concerns   95

Geochemistry and Natural Resources   109

Experimental Techniques and Facilities   117

Data and Sample Accessibility   120

Preface

Orientations in Geochemistry is a review of the current state of the science, and an assessment of the directions in which it will or should move during the next 5 or 10 years. Emphasis is placed on evaluation of areas of greatest potential and on the work that needs to be done to advance knowledge in those areas. The intimate interrelations of geochemistry to many other disciplines, its dependence on them, and its role as an integrating influence among them is a theme that runs through the report and is emphasized in the frontispiece.

The report is addressed to geochemists, to all other practicing scientists and science students, and to administrators of scientific programs. It is also for the use of scientific advisers to groups whose major interests may not be science, but whose decisions must rely in part on the contributions that geochemistry can make to the solution of their problems.

The report was initiated as the result of a request from the National Science Foundation to the U.S. National Committee for Geochemistry of the National Research Council. It was to survey and discuss the major areas of current interest, with emphasis on the most important areas for research and the kinds of work needed for their advancement. Discussions of present and possible applications of geochemistry to problems of immediate concern were requested, especially problems related to the environment and to supplies of natural resources.

In our response to these requests, we began by subdividing the field of geochemistry into four major subject areas—extraterrestrial geochemistry, which consists chiefly of the geochemistry of the solar system; solid earth geochemistry, or the geochemistry generally related to deep-seated processes in the earth; the geochemistry of materials that cycle through the earth's surface environment (exogenic); and organic geochemistry. These subdivisions are far from being mutually exclusive, but serve as the most workable basis for discussion.

The following format for our report finally emerged: An initial summary condenses the major conclusions of the report, and serves as a guide to the chapters that the reader might wish to study at length. An introduction then attempts to cope with the difficult and controversial problem of delimiting the field of geochemistry and gives some of the historical background necessary to understand the scope of geochemistry as recognized today, as well as its complex interplay with many other disciplines. Next come the substantive chapters of the report, covering extraterrestrial, solid earth, exogenic, and organic geochemistry. These are followed by the shorter chapters, "Geochemistry and Environmental Concerns" and "Geochemistry and Natural Resources." Brief technical chapters, "Experimental Techniques and Facilities" and "Data and Sample Accessibility," complete the report.

The scope of geochemistry is so great that any report this short, written by a small panel, will necessarily neglect or minimize the importance of many areas of research or application. Nevertheless, the Panel feels that this report should be a useful first attempt to describe the field of geochemistry in modern terms; it also felt obliged to recommend topics for which additional reports would be worthwhile, such as mineralogy, petrology, and crystallography.

Geochemistry, as we have defined it, is practiced in universities, in many governmental agencies, in research institutes, in the industries concerned with fuels and other material resources, and by private consultants. It is supported by a variety of granting sources, governmental and private. Many geochemical studies are undertaken by scientists who are not identified primarily as geochemists, and who may work in groups not even identified with earth science.

Acknowledgments

This study was performed under the Division of Earth Sciences' U.S. National Committee for Geochemistry by the Panel on Orientations for Geochemistry, which was supported by the National Science Foundation. The Panel wishes to express its appreciation for this interest and support.

The Panel is likewise grateful for the thoughtful reviews by the Committee on Science and Public Policy, the Division of Earth Sciences, the U.S. National Committee for Geochemistry, and the comments by the following individuals: Ernest E. Angino, University of Kansas; Jeffrey L. Bada, Scripps Institution of Oceanography, University of California at San Diego; Hubert L. Barnes, Department of Geosciences, Pennsylvania State University; Paul B. Barton, Jr., Geologic Division, U.S. Geological Survey, Washington; Michael L. Bender, Graduate School of Oceanography, University of Rhode Island; Robert A. Berner, Department of Geology and Geophysics, Yale University; Francis R. Boyd, the Geophysical Laboratory, Carnegie Institution of Washington; Peter R. Buseck, Department of Geology, Arizona State University; A. G. W. Cameron, Harvard College Observatory; Keith E. Chave, Department of Oceanography, University of Hawaii; Felix Chayes, Geophysical Laboratory, Carnegie Institution of Washington; Robert N. Clayton, University of Chicago; Jon J. Connor, U.S. Geological Survey, Denver, Colorado; Jack R. Dymond, School of Oceanography, Oregon State University; P. Edgar Hare, Geophysical Laboratory, Carnegie Institution of Washington; Albrecht W. Hofmann, Department of Terrestrial Magnetism, Carnegie Institution of Washington; Issac R. Kaplan, Department of Geology, University of California at Los Angeles; Klaus Keil, Institute of Meteoritics, University of New Mexico; Dana R. Kester, Graduate School of Oceanography, University of Rhode Island; Kenneth L. King, Lamont-Doherty Geological Observatory, Columbia University; Abraham Lerman, Department of Geological Sciences, Northwestern University; John Lovering, Department of Geology, University of Melbourne; Christopher Martens, Department of Oceanography, Florida State University; Brian Mason, Department of Mineral Sciences, Smithsonian Institution; Warren G. Meinschein, Department of Geology, University of Indiana; Carleton B. Moore, Center for Meteorite Studies, Arizona State University; Karlis Muehlenbachs, Geophysical Laboratory, Carnegie Institution of Washington; John A. O'Keefe, Goddard Space Flight Center, NASA; Pierre L. Parker, Institute for Marine Science, University of Texas; James G. Quinn, Graduate School of Oceanography, University of Rhode Island; Charles Schnetzler, Goddard Space Flight Center, NASA; Brian J. Skinner, Department of Geology and Geophysics, Yale University; James V. Smith, University of Chicago; George R. Tilton, University of California at Santa Barbara; Karl Turekian, Department of Geology and Geophysics, Yale University; John W. Winchester, Department of Oceanography, Florida State University; and Hatten S. Yoder, Jr., Geophysical Laboratory, Carnegie Institution of Washington.

The Panel gratefully acknowledges the artistic help of Mary-Hill French, Stanley M. Hart, A. Conrad Neumann, John W. Larimer, Donald W. Joyce, and James M. Gormley with the illustrations.

Summary

Geochemistry is discussed in four major chapters, "Extraterrestrial Geochemistry," "Solid Earth Geochemistry," "Exogenic or Low-Temperature Aqueous Geochemistry" (essentially surface processes and cycles), and "Organic Geochemistry." In addition there are the short chapters, "Geochemistry and Environmental Concerns," "Geochemistry and Natural Resources," "Experimental Techniques and Facilities," and "Data and Sample Accessibility." An assessment is attempted of the major types of work now in progress within each of those subdivisions, and suggestions are offered concerning future work that should be done. Only a brief account is given of some of the broader aspects of these recommendations.

All aspects of geochemistry are entering a stage of extensive numerical simulation of natural processes, and modeling of various systems will expand markedly, with an accompanying demand for extensive quantitative analytical data. This demand, in turn, will necessitate expanded use of existing types of analytical equipment and the development of new types and techniques. A further requirement will be the development of data banks and better data-retrieval systems, as well as coordinated efforts to provide standard samples of natural materials for interlaboratory comparison.

The increasing emphasis on tracing elements and nuclides through their cycles is placing geochemistry in the role of a science that utilizes the results of many other disciplines, making extensive use of the data of such currently separate areas of investigation as marine chemistry and biology, atmospheric sciences, soil science, hydrology, and ore deposits. Geochemistry, in its turn, provides information to these allied fields; for instance, in its studies of nutrient additions to the oceans and their controls.

In terrestrial geochemistry, a major theme emphasizes investigating in detail the transfer of materials in all earth environments. The solutions to many current problems, especially those of environmental pollution, and the development of new mineral resources, will require knowledge of sources, methods of transfer, and sinks for many elements.

Our survey showed that present knowledge of the sources and sinks of many elements is fragmentary, and that strong efforts should be made to supply the missing information. Only through the detailed investigation of the natural earth system, which has been operating dynamically for hundreds of millions of years, will it be possible to make satisfactory predictions of the effects of man's additions to the exogenic cycle. The chapter on the environment emphasizes this point. The same is true for any endeavors to assess the reserves of mineral deposits. The hypothesis that the earth's crust comprises a series of plates that move relative to each other has already led to new ideas as to the origin of metalliferous deposits and might be a guide to the discovery of important reserves.

Great progress has been made in laboratory simulation of deep-earth conditions. A continuing goal is to develop experimental techniques that will permit duplication of earth conditions to the core itself. This will also demand continued development of chemical theory to interpret the experimental results. The recent development of experimental apparatus that permits investigation of the behavior of volatile substances at high temperatures and pressures is beginning to close the gap between experimental work on metalliferous deposits and its use in exploration and is providing general insight into deep-earth processes.

Extraterrestrial geochemistry has a major task in analyzing in detail the materials recently made available to it. Nonetheless, it has kept pace remarkably well in interpretation of the rapidly accumulating data. We now know that the earth, moon, and meteorites all formed within a hundred million years or so (about 4.7 billion years ago), and that all bodies in the solar system did not have the same primordial composition. Future work will involve a continued major analytical effort and a strong effort to work out, in cooperation with astronomers, the details of the origin of the solar system.

Organic geochemistry also must continue to use the new analytical tools that have revolutionized the field. One major goal is to understand the changes in terrestrial organic materials with increasing geologic age, so that as much as possible can be interpreted of the record of life through time, especially the critical stage of origin. Another goal is expanded research on extraterrestrial organic materials. Recent discoveries have revealed the presence of organic compounds of nonbiologic origin in meteorites.

Earth and planetary science appears to have reached a stage at which the capability of understanding the earth in detail as a dynamic chemical system has been achieved. In our survey of geochemistry, we were appalled by the present ignorance of what should be known about our stressed planet, and by the obvious absence of various kinds of basic data. Most of the chapters are chronicles of ignorance; almost every paragraph complains about data that are obtainable but unobtained. On the positive side, material and energy balances involving the whole solar system are not beyond our grasp. The next decade should see great progress in the deciphering of the history of the solar system, the origin of life, and the metabolism of the earth.

Recommendations for Development of Information Relative to Geochemistry Beyond the Scope of this Report

Because of limitations of time and money, several aspects of geochemistry that should be investigated could not be treated adequately in this report. Important among these are the following:

1. The current size of the national and international geochemical effort, in terms of man-years and dollars or dollar equivalents invested. We were unable to assess accurately the size of the United States effort, and could make no attempt to compare it with the effort in other countries. Russia, in particular, probably spends far more man-years per year than does the United States, but this estimate is based qualitatively on relative numbers of publications in the field. A partial survey of United States funding sources indicates that a minimum of $30 million, and an estimated total of about $50 million, are spent.

2. Environmental geochemistry and geochemistry as applied to

natural resources. These are subjects that we have treated only in the most general terms. We suggest that environmental geochemistry would be worthy of a much more detailed report. One aspect of environmental geochemistry, *Geochemical Environment in Relation to Health and Disease,* is the subject of a recent lengthy report. Geochemical aspects of the natural resources problem also are being studied by many groups. Material for an integrative report on the specific role of geochemistry as applied to natural resources is available.

3. The importance of modern developments in instrumentation. This topic has been stressed in this report, but only in the most general terms. Details are needed of the many types of analytical techniques available today, their potentials, limits, costs, and their current and future capabilities.

4. Problems related to the collection, analysis, and maintenance of standard samples of geochemical materials, and those of integration of the unpublished chemical analyses of natural materials that exist throughout the international scientific world. It is likely that the data required to solve a number of the problems listed in this summary exist today, but are unavailable because of lack of effort to unearth them.

5. The problem of information storage, retrieval and use. This problem is perhaps the most distressing. Especially in geochemistry, which draws on, or should draw on, the data and concepts of many other fields, it is impossible for an individual scientist, even if he could find all the information relevant to his problems, to assimilate and use it. The Panel decided that the solution of this problem was critical to progress, but felt that a separate study was necessary.

6. The geochemical aspects of mineralogy, petrology, and crystallography. Treatment of these aspects is too limited in this report. The separation of these areas from geochemistry is most difficult, and there is no general agreement on how to do it. A separate report on those fields, including their roles in geochemistry, would be useful.

Orientations in Geochemistry

Introduction

Geochemistry has been growing rapidly, developing new fields, impinging on the domains of classical disciplines, interpenetrating new ones, sometimes parasitically, sometimes symbiotically. Today it covers a wide range of subject material and uses analytical data covering almost every aspect of the chemistry of the universe. Geochemists themselves obtain many of the data, but they also use results obtained by chemists, biologists, physicists, astronomers, medical scientists, atmospheric scientists, environmental engineers, and many others. Our best definition is that *geochemistry is the area of science concerned with the chemical composition and the chemical evolution of the earth and solar system, and it includes the chemical aspects of all the sciences that contribute to it.*

The interweaving of geochemistry with other sciences had made it more and more difficult to separate as a distinct entity and has broadened the scope of activities now classified as geochemistry.

Geochemistry includes the chemistry of the cosmic dust from which the solar system was formed; the chemistry of the accreting earth, moon, and planets; the chemistry of the earth's crust, mantle, and core; the chemistry of the rock cycle, involving erosion, transport, deposition, and uplift; the chemical evolution of the oceans

1

and atmosphere; the chemistry of organic materials in rocks. Thus *all chemistry involved in the context of earth and planetary evolution is geochemistry.*

The kind of difficulty that rises in delimiting geochemistry from other fields is illustrated by the question of the degree to which an article that describes the morphology, crystal structure, and chemical composition of a mineral should be classified as mineralogy or geochemistry. A generation ago, such an article would have been classified entirely as mineralogy. Today it would be classified in part or even completely as geochemistry, particularly if there were aspects of the report devoted to the thermodynamic properties of the mineral, to its phase relations to other minerals, or to the chemical environments in which it could be formed. To some extent, geochemistry is growing by acquisition of parts of other disciplines, while at the same time other disciplines have tended to expand to include additional geochemical aspects. As the boundaries between classical earth-science disciplines become less distinct, more of the gray areas are being called geochemistry.

Perhaps the closest parallel to geochemistry is organic evolution, for which the interpretation of changes in organisms with time depends on genetics, molecular biology, biochemistry, ecology, agriculture, and other fields that are in themselves basically time-independent.

To put geochemistry into its modern framework, the best device may be a brief history.

HISTORICAL DEVELOPMENT

The roots of geochemistry extend far into the past, but before about 1900, chemistry and natural sciences were closely related, and a great deal of geochemical work was done on the composition and evolution of rocks and minerals and natural waters by scientists of various callings. Classical examples are the work of Sorby on the chemistry of fluid inclusions in minerals around 1860 and the work of Van't Hoff on the phase relations in the Stassfurt salt deposits of Germany about 1900. Sorby was classified as a geologist and microscopist; Van't Hoff occupied at various times chairs of physics, chemistry, mineralogy, and geology.

For the next 40 years or so, there was a marked parting of the ways, with geologists turning toward areal mapping and the deciphering of the history of the earth from the interrelations of fossils, rock layering, metamorphism, and intrusion of igneous masses, while chemists abandoned minerals for the more easily controlled studies of pure compounds. Geochemistry was not neglected entirely, and magnificent work was continued by scientists such as F. W. Clarke in the United States, V. M. Goldschmidt in Germany, V. I. Vernadsky in the U.S.S.R. Emission spectroscopy was one of their major tools. In the United States, some laboratory studies were carried out in the laboratories of federal agencies, such as the U.S. Geological Survey, and in private institutions, such as the Geophysical Laboratory of the Carnegie Institution of Washington, and, on a small scale, in various universities. Both the mining and fuel industries were active. The major efforts in the geological sciences were the deciphering of earth history from the field relations of rocks, and the identification, associations, and depositional sequences of minerals.

A major advance came in the 1930's in the widespread application of X-ray diffraction studies to the deciphering of the structures of minerals, which permitted determination of the phases present in fine-grained rocks that had defied the ordinary polarizing light microscope. For the first time, information was available on the mineral content of shaly rocks, which until then had been simply called "clay." At that time, also, the mass spectrograph was developed to the stage where it could be used for the determination of the absolute ages of rocks by measuring isotope abundances, but little work of this nature was done. The potential also existed for studies of stable isotopes.

For whatever reasons, the real potential of X-ray methods and isotope determinations, as well as of other instruments and techniques that had already been developed, did not begin to be realized until the late 1940's and early 1950's. Perhaps it was only the characteristic 20-year gap between invention and widespread application, but in part the geochemical renaissance probably resulted from the beginning of federal funding of university research in the late 1940's and early 1950's. After World War II, there was also a ferment among geologists, who began to realize that the high percentage of total effort devoted to mapping was out of balance and that new concepts could arise from use of new laboratory equipment and techniques. At any rate, in the United States the designation of a professional scientist as a geochemist postdates 1950. The Geochemical Society was not founded until 1955.

In the late 1940's, 1950's, and 1960's, geochemical laboratories sprang up in universities all over the United States. Initially emphasis was on high-temperature, high-pressure studies of silicate equilibria, following the long tradition of productive experimentation at the Geophysical Laboratory, Carnegie Institution of Washington, but radioactive age dating and measurement of stable isotope ratios were also being rapidly developed. The major federal effort to develop uranium deposits in the United States during the late 1940's and early 1950's required application of sophisticated geochemistry to interpret the origin of uranium ores, and thus guide exploration and mining. The program, which was markedly successful, gave geochemistry a big push.

In the 1960's, with strong federal support for space and ocean exploration, geochemistry grew exponentially. Methods of measuring the absolute ages of rocks proliferated; for the first time a detailed chronology of earth history, trustworthy in terms of the actual range of years before the present in which various events occurred, could be established. Studies of stable isotopes also proved highly rewarding: For example, major variations in sulfur isotope ratios of the oceans through time were discovered; isotope ratios were used to measure the temperatures of ancient seas and to trace the genesis of underground waters. Moon samples were studied mineralogically and chemically, chiefly by scientists trained in earth sciences, so that lunar rock studies became closely identified with geochemistry.

Once studies of the chemical aspects of geology got under way, and students adequately trained in basic sciences were produced, the half-century-long divorce of geology and chemistry was ended. The reunion has been remarkably fruitful. The theories developed in chemistry during the hiatus between the disciplines were rapidly applied to studies of geologic processes. Theories of electrolytes developed in chemistry 20 to 30 years before were applied, especially in seawater chemistry. Thermodynamic data for minerals, completely lacking in 1950, have now been obtained for many of the important species, and provide a predictive framework for interpretation of experimental and observed mineral-phase relations. Geoscientists found that analytical techniques developed in chemistry were the answer to many of their problems requiring quick and accurate analyses of endless samples needed to characterize the composition of the rocks and waters and atmospheres of the earth and planets. In some areas, geochemists adapted and extended such techniques; they became leaders in developing precise analytical instruments.

This is especially true in some recent projects designed to explore the chemistry of the oceans, where information is needed on the details of the lateral and vertical chemistry of a water body 4 km deep covering 70 percent of the globe. Geochemists have even outstripped chemists in some areas of studies of electrolytes, such as in the complex concentrated solutions represented by natural brines and in determination of ionic interactions in aqueous solutions at high temperatures and pressures. They are also in the forefront of the application of thermodynamics to polyphase systems, and in the graphical and mathematical descriptions of such systems. Their development of apparatus for studies of extremely high temperatures and pressures, impelled by their interest in the chemistry of the deep interiors of planetary bodies, has kept them among the leaders of this general field.

Geochemists responded rapidly to the development of paper and gas chromatography for the study of complex natural organic compounds. Some of their work in studying the organic compounds of ancient rocks, meteorites, and lunar samples required the ultimate in refined "clean-laboratory" techniques and ultramicroanalytical methods. They applied ion-sensitive electrodes to field and laboratory studies of natural aqueous systems and immediately saw the usefulness of atomic absorption analysis. Geochemists are deeply involved in the use and development of remote-sensing devices, for determining rock composition by spectral response, for example.

Another major direction of geochemical growth, already discussed in part, has evolved from the development of instruments and techniques particularly useful in solving the classical analytical problems that are posed by natural chemical systems. In the last few years, the electron microprobe and the scanning electron microscope have been extensively used by geochemists. With the microprobe, it has been possible to determine the details of the composition of coexisting mineral phases and, for the first time, to estimate the degree to which equilibrium is attained in complex mineral systems. The scanning electron microscope permits direct checks on predicted alteration products of minerals. A whole new field of study was created in which the surface characteristics of mineral grains are used to decipher their chemical and physical history. For example, with high magnifications it is possible to see the pitting of quartz grains resulting from wind transport, modified by later solution processes, and thus to trace their depositional history.

Low-temperature experiments attempting to synthesize silicate minerals from aqueous systems commonly produce products for which X-ray analysis shows no structure, but infrared spectrometry has permitted determination of chemical bonding otherwise invisible.

The use of the computer in the analysis of crystal structures, with its nearly automatic resolution of electron-density data into atomic configurations, has made it possible to work out the structures of minerals of complex chemical composition in a relatively short time, and the void in knowledge of the detailed structures of minerals, paralleling the void in knowldege of their thermodynamic properties, is rapidly being filled.

Neutron activation analysis is another technique that has been solving hitherto impossible analytical problems for geochemists. It has permitted determination of incredibly low concentrations of elements, in the fraction of a part-per-billion range, and generated new hypotheses of rock genesis that were not possible from the older analyses that were incapable of "seeing," much less separating, many elements in extremely low concentrations.

Many other examples of evolving new analytical techniques are given in the text that follows. Their influence has been to accelerate remarkably the development of geochemistry. As the attention of earth scientists turned to chemical problems, they were almost overwhelmed by the rate at which they were presented with the ability to measure variables that were beyond their reach only a few years ago.

THE WEDDING OF THE NATURAL SCIENCES

Geochemistry, with its emphasis on time-dependent processes, has established a peculiar position. Its practitioners are relatively few, and its total budget small relative to the money spent on obtaining the data used. Yet it is a strong force in pulling together many other disciplines. Geochemistry is concerned with the movement of elements, sometimes cyclic, sometimes not; because it traces pathways, sources, and sinks, it tends to unify these particular aspects of many other disciplines.

Soil scientists, for example, have been interested in soil productivity, the relation of soil types to rock types and climates, the effects of the addition of fertilizers, the soil atmosphere, and the minerals of soils and their interrelations, but they have made few studies of the material balances of soils, rain, percolating waters, and their influence on groundwater composition and the transport of materials from soils to streams. Geochemists, with their interest in element cycling, have tended to unite studies of atmospheric composition and precipitation to element migration and phase transformations in soils, and thence to the compositions of underground waters and their relations to those of rivers and lakes.

This difference in point of view has caused fertile interactions with atmospheric scientists, soil scientists, chemical hydrologists, and water quality people. The geochemical approach has been to regard soils as reactive multiphase systems; collaboration with soil scientists, who have emphasized the ion-exchange properties of soils, has resulted in new concepts useful to both groups.

Similarly, the preoccupation of geochemists with the mineral sources of dissolved solids in streams has focused the attention of water resource scientists on the role of mineral–water reactions in controlling water quality. A generation ago, it was almost impossible to obtain water analyses accompanied by data on the mineralogy and chemical composition of the rocks from which the waters were derived; today, studies of the interrelation of water composition and solid source materials are numerous.

Studies by geochemists of the sources and sinks of dissolved materials entering the oceans have been responsible for major changes in the scope of marine chemistry. Today there is a strong new interest in the oceans as a heterogeneous system with continuously dissolving and precipitating materials, with some major reactions dominated by organisms, others not. The results of chemical studies of sediments from the bottom of the deep oceans have demonstrated the important role of postdepositional chemical processes in controlling the composition of seafloor sediments.

Interest in element cycles and material balances has also helped to create interest in the interface between ocean and atmosphere (thus uniting atmospheric science and oceanography) and the interface between ocean and bottom sediments (thus uniting chemical oceanography, geochemistry, sedimentation, stratigraphy, and paleontology).

Astronomy and geology have always run hand in hand, as exemplified by the famous collaboration of T. C. Chamberlain, a geologist, and F. R. Moulton, an astronomer, in developing, a half century ago, the planetesimal hypothesis of the origin of the solar system. Today, the collaboration between astronomers and geochemists includes continued efforts to understand the origin of the earth and solar system, as well as the origin of the elements themselves.

As geochemists developed and used new analytical techniques, including methods of age-dating with radioactive isotopes and the measurement of the rates of loss of optical activity of amino acids, they also began to contribute to archeology and anthropology. They showed that ^{14}C dating of archeological sites could be used to provide archeologists with an absolute time scale. Their experience with microanalysis of compounds aided identification of the ages and sources of objects such as old coins, statues, and paintings. A surprising number of geochemists are developing at least a part-time interest in archeology, art history, and similar studies.

More examples could be given of the interweaving of geochemistry with other sciences, but perhaps enough has been said to illustrate the earlier statement about its parasitic and symbiotic relations to many other classically defined fields. The current effort in geochemistry is small compared to many other scientific disciplines, but geochemistry is having a significant effect on them by raising new questions, demanding new data, and encouraging communication among the other disciplines.

At present, a major expansion of geochemistry is in the area of health and disease. Geochemists have been leaders in the accumulation of analyses of trace elements in rocks, waters, and plants, initially motivated by attempts to use such analyses to find mineral deposits (e.g., selenium-concentrating plants were found to be guides to uranium deposits in the western United States). Work of this kind has led to studies in populated areas of the effects of local rock types and vegetation on the incidence of various diseases. These studies led to the discovery that the rock types on which people live control the levels of their exposure to natural radioactivity and to the expansion of information on the relation of the trace element content of plants and animals to the composition of the bedrock on which they grow.

As contamination of air and water by a multitude of substances was investigated, initially by scientists of various kinds in a more or less random fashion, it became clear that the establishment of baselines for pollution—the preman circulation of elements in the earth-surface cycles, with their sources and sinks—was a prerequisite for assessment of man's interference. The limited data initially available for constructing such baselines was in the hands of the geochemists. Some geochemists have turned their research efforts completely in the direction of gathering more data for the baselines; others have altered the emphasis of work already in progress to include acquisition of pertinent baseline data.

Another application of geochemical information derived from studies of the circulation of the elements is helping to find additional sources of economic mineral deposits. Unusual degrees of concentration and ease of access are the required aspects of mineable deposits. More and more reliance must be placed on lower grade material or discovery of new high-grade ores in hitherto unsuspected sites. The greater the knowledge of the details and controls of element cycling, the better the chances for effective future development of the most economic sources.

The rapid growth of geochemistry, with new directions of research rapidly raising new problems, has revealed marked imbalances in its former and present interests. Much of the growth has been in extension and ramification of directions popular at the beginning of expansion, but which are not providing the data necessary to solve the new problems. In retrospect, it is hard to understand the reasons for the neglect of many areas of research that were recognized many years ago as important for the rounded development of geochemistry.

Much of the ensuing discussion is devoted to revealing these information gaps. We hope that this report will stimulate additional consideration of the research goals necessary for maximum progress.

Extraterrestrial Geochemistry

DEFINITIONS AND GOALS

Extraterrestrial geochemistry is the study of the chemical history of nonterrestrial materials, using as its primary sources of data meteorites, lunar samples, and astronomical observations. The field grew out of the mutual interests of astronomers, geologists, and chemists in solving classical problems regarding the age, composition, and origin of the earth. These remain the primary goals, but the data and ideas that have been generated bear on additional broad scientific problems ranging from nucleosynthesis to the origin of life.

RECENT ADVANCES

The most spectacular advances in recent years have resulted from the study of lunar samples. We have learned that the formation of the earth, the moon, and the meteorites were nearly simultaneous events that occurred approximately 4.5 billion years ago, supporting the concept that the solar system as a whole formed at this time. The cessation of major volcanic activity on the moon 3 billion years ago contrasts, however, with the continued activity on the earth. In addition, the moon appears to be relatively poor in volatiles, such as water and alkalies (sodium and potassium), as well as noble metals such as gold and palladium. The refractory elements aluminum, calcium, and titanium are relatively more abundant than they are in meteorites or on the earth. These compositional differences indicate that the moon, and by implication, other bodies within the solar system, do not all possess solar or "primitive" composition.

Although they have been partly eclipsed recently by lunar studies, meteorites must still be regarded as the "Rosetta stones" of the solar system. Moon rocks provide information chiefly about primitive geologic processes; whereas meteorites, being more pristine, contain a record of the very early solar system and how it formed. Lunar and meteorite research should not be regarded as competitive, however, because the two fields are inextricably linked. As an example, some current theories on the bulk composition and accretion of the moon rely heavily on the Allende meteorite that fortuitously fell in February 1969, a few months before the flight of the Apollo 11, the first manned landing on the moon. This stone contains fragments, enriched in refractories and depleted in volatiles with respect to cosmic abundances, that may represent the original condensates of the solar nebula.

The major recent advances in meteoritics reflect the interdisciplinary nature of the field. They include the discovery of extinct radioactive elements that indicate meteorites formed within 100 million years after the elements themselves. These studies advance understanding of nuclear processes in stars and serve as a precise dating method for events that occurred in the very early solar system. Impressive advances have also been made in understanding the condensation of particles in the cooling nebula and their accretion into planetary bodies. These studies, based largely on meteorites, appear to provide a useful framework for a major synthesis of astronomical and geochemical data. The postaccretion thermal history of meteorites is now fairly well understood, including their metallurgically based cooling rates; although the nature of the heat source remains a mystery. Organic chemists, after some frustrating failures, have now shown that amino acids, the basic building blocks of life, are present in some meteorites and are of nonbiological origin. This discovery obviously has broad implications with regard to the origin of life in our solar system.

FUTURE EFFORT

Extraterrestrial geochemistry can now begin to assemble data from several sources in the solar system and use these to construct unified theories that involve both the formation of the bodies of the solar system and the time scale of formation, as well as the processes and time scale of the differentiation of the planetary bodies. The following important and more specific areas appear to be worthy of investigation in the future:

• The resolution of the time scale. This requires the application of recently refined and developed age-determination techniques to resolve which meteorites formed first and to determine the total accretion time scale. These results, in turn, must be related to what has been learned about ages of the earth–moon system.

• The condensation and fractionation processes that took place before and during accretion in the solar nebula. These studies will be based on experimental and theoretical data on the condensation of minor and trace elements and key compounds, consideration of the kinetic problems attending vapor condensation, and analytical and mineralogical studies of high-temperature inclusions in meteorites.

• The primitive differentiation processes that operated on the moon and in meteorite parent bodies. These studies will require continued laboratory experimentation on pertinent synthetic systems, analytical determination of trace element abundances in individual phases, and an evaluation of the distribution coefficients under various pressure–temperature conditions. Although these research interests correspond to similar interests in the field of solid earth geochemistry, the phases and thermodynamic conditions of interest in these subdisciplines are not the same.

• Origin and stability of organic matter. A clearer understanding of the conditions necessary for synthesis requires continued laboratory experimentation. These studies, in which observed and synthesized molecules are compared, should soon include carbon isotopic ratio determinations. Relations between interstellar molecules and meteoritic organic compounds are unexplored but will be of mutual interest to astronomy and geochemistry.

To some extent, progress in extraterrestrial geochemistry will depend on analysis of other bodies in our solar system, which will involve the geochemist in space missions. The greatest progress will come when samples are returned for investigation in terrestrial laboratories.

METEORITE RESEARCH

Meteorites (Figure 1) are the oldest accessible materials, with ages of up to 4.7 billion years—the generally accepted age of the solar system. The most primitive meteorites, chondrites, have long served as the basis of the "cosmic-abundances" curve on which models of nucleosynthesis are based and tested. Meteorites are pieces of asteroids and comets, broken off by collisions among these bodies in their journeys through the solar system. They provide geochemists with samples of the deep interiors of bodies to which they otherwise would not have access. These pieces are often quite small, which makes analysis a difficult and demanding job, forcing the geochemist to perfect existing analytical techniques or to invent new ones. This technological spin-off should not be overlooked in evaluating the contributions of meteorite research.

For clarity, the following discussion is divided into three broad sections: ages, elemental and isotopic composition, and mineralogy and texture. A separate section on extraterrestrial organic matter at the end of this chapter includes a discussion of carbon chemistry in meteorites.

Ages

A variety of events in a meteorite's history can be dated. In chronological order, these are (1) the time between the end of nucleosynthesis and formation as a solid body, (2) the age of the sample and its residence time in a parent body, (3) the time of breakup of parent bodies, and (4) the time the meteorite has been exposed to cosmic radiation before arriving on earth.

Extinct Nuclides Since the discovery of excess ^{129}Xe in chondrites, attributed to the decay of extinct ^{129}I [half-life $(T_{1/2}) = 17$ million years (m.y.)], considerable effort has been expended in the study of xenon isotopes. In the late 1960's, it was proved that part of the excess of heavy Xe isotopes (134 and 136) was caused by the decay of ^{244}Pu $(T_{1/2} = 80$ m.y.). The amount of ^{129}Xe, ^{134}Xe, and ^{136}Xe found in meteorites, can thus be used to date the time between the end of

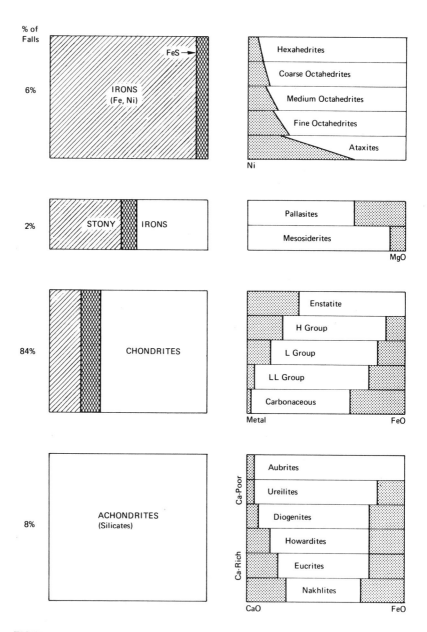

FIGURE 1 There are four broad classes of meteorites (left) each with a characteristic metal-to-silicate ratio (shaded and white, respectively). Each class is further subdivided based on various textural features and compositional criteria, such as the content of nickel, magnesia, metallic iron, iron oxide, and calcium oxide.

nucleosynthesis, when ^{129}I and ^{244}Pu ceased to form, and the time the meteorites began retaining xenon. These data indicate that the last nucleosynthetic event occurred about 100 million years before the separation of the solar nebula from the interstellar medium. Theoretical studies on the importance of a prompt nucleosynthetic event at an early stage, followed by a long period of continuous synthesis, and concluding in a last-minute event have been attempted. But such studies are still too model-dependent to yield definitive information. Continued efforts are required to delimit the initial plutonium/uranium ratio through fission-track and xenon-isotope studies.

The short half-life of these nuclides is advantageous for resolving events during the formation of the solar system. On the basis of data from 17 meteorites, representing a wide variety of types, it appears that most began retaining xenon within 6 to 8 million years of each other. Some of this difference appears to be real, as the error for individual cases is only ± 1 million years. However, more comprehensive data obviously are required.

In some chondrites, enrichment of the heavier isotopes of xenon (131–136) is observed, suggestive of heavy-element fission, but the observed ratios do not correspond to any known nuclide. It has therefore been suggested that the radionuclide may have been a superheavy element with an atomic number (Z) of from 112 to 119. The possibility is intriguing because an "island of stability," centered on $Z = 114$, A (atomic weight) = 298, has been predicted from nuclear systematics.

Rubidium/Strontium   Some spectacular advances in the rubidium/ strontium age-dating method over the past 10 years permit much more precise dating than has heretofore been possible. In some instances, resolution of events separated by ± 2 million years can now be made. The key result has been that, with a few interesting exceptions, all classes of meteorites, including the differentiated irons and and igneous achondrites, formed within 100 million years of each other.

Potassium/Argon   A new method, capable of providing many new insights, has recently been developed. A sample is first irradiated to convert ^{39}K to ^{39}Ar, then heated, usually at 100-degree intervals, and the released gas is analyzed in a mass spectrometer. At each temperature interval, an age is determined. Poorly retentive sites release their gas at low temperatures; the resultant ages are normally younger than the whole rock age, as is to be expected. This method can be used to date rocks that have experienced a partial gas loss during metamorphism or impact. The method has proved itself in the study of lunar samples, but it has yet to be widely applied to the study of meteorites, although results have been obtained that confirm that the L-Group chondrite parent body broke up about 500 million years ago.

Uranium/Lead Refined analytical techniques now permit more precise measurements of the lead isotopes in meteorites. Accurate estimates of the primordial lead isotopic ratios are important, being used to estimate the age of the earth as well as of meteorites. The uranium/lead method also has the potential of being used to resolve events in the early solar system that are separated by about 5 million years. With the combined information from uranium/lead, rubidium/strontium, and xenon studies, it should be possible within the next few years to answer questions such as which type of meteorite formed first, what is the total accretion time scale, and what, if any, time lapse occurred between the formation of the meteorites and that of the earth and moon.

Cosmic-Ray Studies Meteorites are exposed to cosmic rays when they are broken from their parent bodies. The interaction of the cosmic rays with the meteorite induces nuclear reactions leading to the production of radioactive isotopes and excess quantities of noble gases. Analysis of the amounts of these cosmic-ray produced materials can be used to estimate the meteorite's exposure age as small bodies. Clusters of exposure ages for groups of meteorites indicate that common collisions released many specimens. For example, clusters exist for H-Group chondrites (4 m.y.), aubrites, achondrites containing enstatite (34 m.y.), and diogenites, achondritic meteorites composed essentially of orthopyroxene (13 m.y.).

Cosmic rays, as well as particles from the solar wind and solar-flares, burrow into exposed materials creating a track. The number of such tracks is related to the exposure age and depth within the sample; whereas the length of the track is related to the energy of the ray. Studies of these phenomena can thus be used to estimate the preatmospheric size of the meteorite and the time and depth of exposure, and to test the theory of constant cosmic flux in time and space.

Fission Tracks Atoms produced during fission (of uranium or plutonium) also leave tracks. The number of these tracks can be used to date samples, estimate the plutonium/uranium ratio, and even determine cooling rates, because each mineral begins to register tracks at a unique temperature. Track studies thus provide a variety of kinds of information regarding both the history of the meteorites and the history of the solar system.

Elemental and Isotopic Composition

The 1950's and 1960's witnessed a concerted effort to determine the composition of meteorites. The major advance was in the determination of trace elements by neutron-activation analysis, a technique capable of obtaining accurate analyses in the part-per-billion range. Comprehensive models based on trace element data are now possible. Future studies can be aimed at key elements to test various models.

Volatile Elements   In his classic papers 20 years ago, Urey pointed out how volatile elements could be used as cosmothermometers to estimate accretion temperatures. These ideas have recently been updated using newer data. Accretion temperatures based on lead, bismuth, indium, and thallium have been determined for about 100 ordinary chondrites. The results are generally consistent and point to temperatures of about $450 \pm 50°K$. These studies have important implications with regard to the temperature–pressure relations that existed in the nebula during accretion. The present blackbody temperature in the region of the asteroid belt is only $170°K$. This implies a transient heat source during the accretion of the meteorites. One possibility is the Hayashi stage (an early, highly radiant condition) of the sun; another is the gravitational energy released during the collapse stage of the solar nebula. A potentially fruitful area of interest to astronomy and meteoritics is a comparison of the time–pressure-temperature relations obtained from theoretical and observational studies of protostars with evidence from meteorites.

Further studies of volatile elements will be of considerable interest. It will be important, for example, to explore the effect of redistribution of the volatile elements during metamorphism, the relative importance of nonequilibrium during condensation, and the possibility that the nebular gas was partly ionized.

$^{18}O/^{16}O$ A model of the distribution of oxygen isotopes during condensation in the nebula has recently been developed. The results indicate that ordinary chondrite material became separated from the cosmic gas over the temperature range 500–$400°K$. These results are in good accord with the trace element data, but additional studies in which both trace elements and oxygen isotopes are measured on the same samples should provide a more definitive test.

Estimates of the accretion temperatures for the earth and moon have also been made. The temperatures (T) inferred from volatile trace element content and $^{18}O/^{16}O$ ratios agree fairly well. Both thermometers point to T $\sim$ 450°K for the earth; whereas for the moon the volatile elements indicate T $\sim$ 500°K and $^{18}O/^{16}O$ ratios indicate T $\sim$ 450°K.

Oxygen isotopes have also been employed to determine metamorphic temperatures in the meteorite parent bodies. These results are broadly consistent with the mineralogical thermometers and provide important clues to the thermal history of planetary bodies.

Refractory Elements The first materials to condense from a gas of cosmic composition have been predicted to be rich in calcium, aluminum, titanium, and rare earths. Interest in these predictions has been considerable for several reasons: Fragments with the predicted mineralogy and composition have been discovered in some carbonaceous chondrites; the earth and moon may be enriched in this early condensate; interstellar gas is depleted in these elements compared to more volatile elements; and most chondrites, excluding the carbonaceous ones, are depleted in highly refractory elements. Evidently, an important event occurred during the early stages of the condensation process, which led to a segregation of the initial condensate and widespread movement of the segregated material. Studies of interstellar gas suggest such events may be important, not only to our solar system, but to the genesis of stars in general. If the inclusions in carbonaceous chondrites represent the initial condensate, studies of their age, composition, and mineralogy should be especially fruitful. Additional theoretical and experimental studies of the exact conditions of formation also seem necessary. The question of how this material came to be enriched in the earth and moon is speculative but deserves further thought.

Noble Gases In some meteorites, the noble gases are so enriched that cosmogenic and radiogenic origins must be ruled out. These excesses have been attributed to solar and planetary sources. Solar gas is found sporadically in all types of meteorites. It is commonly associated with shock and brecciation features and appears to represent trapped solar-wind gases. The possibility that the trapping took place before accretion has also been suggested but remains speculative.

Planetary gas is probably of more ancient origin. Assumed to have been incorporated in the meteorites at the time of their origin, it is more enriched in very primitive meteorites, carbonaceous chondrites and unequilibrated ordinary chondrites. If this gas was incorporated at the time of origin, three aspects need to be explained: total amounts, elemental ratios, and isotopic ratios. A number of mechanisms have been proposed to explain the detailed observations. These include: equilibrium solubility, absorption and grain growth, exotic cosmogenic reactions and thermal diffusion.

Resolving the question of which mechanisms have been important will require studies of several types. Laboratory studies of the solubility of the gases in meteoritic minerals are obviously required. Also, additional information on the mineralogic sites of the gases in conjunction with the appropriate experimental studies will be necessary.

Iron Meteorites Over the last 20 years, the composition, texture and mineralogy of iron meteorites have been extensively studied. Some significant groupings on the basis of texture, trace element content and cooling rate have been revealed. However, there has been no comprehensive explanation of how and why the differences between the groups arose nor of the significance of the trends within the groups.

The central question now is how, when, and where these groupings were established. Do they imply numerous parent bodies? Or perhaps a number of iron bodies distributed throughout a few parent bodies? There is also the question of whether the determining step in the trace element content was composition of the parent body or the pressure (P)–temperature (T)–oxygen partial pressure (pO_2) conditions under which the metal segregated from the silicate. The analytical data appear to be sufficiently complete to begin developing a comprehensive model. Experimental studies of the distribution of trace elements between metal and silicate at various T–pO_2 conditions should prove fruitful.

Stony Irons These meteorites are interesting because, in principle, they should not exist. Silicate and metal should segregate gravitationally in a relatively short time. The mineralogical and textural relations in these meteorites have so far only been surveyed; their compositions and ages have not been extensively studied but should be. Cooling rates of both the pallasites (containing metallic iron and olivine) and mesosiderites (containing metallic iron, plagioclase and

and pyroxene) are the slowest yet determined for meteorites, about 0.1 degree per million years, for reasons that remain unclear.

The possibility that mesosiderites are unusual products of the original condensation has been suggested. If this is so, their ages, noble gas contents, ^{129}Xe contents, and the presence of fission tracks attributable to ^{244}Pu should be of considerable interest. Their bulk composition could be similar to that of the earth. If so, then these rare but interesting meteorites could prove fruitful in providing new insights into the accretion and early evolution of planetary bodies.

Chemistry of Interstellar Grains Astronomers have shown a growing interest in the study of interstellar grains, a field that promises to provide yet another link between geochemists and astronomers. The absorption and scattering characteristics of interstellar grains have been measured in the ultraviolet and infrared. At present the data cannot unambiguously be translated into chemical compositions, although a variety of suggestions has been made; including graphite grains, ice, silicate particles, metal grains, and ice- or graphite-coated grains. A measurement that may provide additional information is the polarization of starlight; however, these results are also ambiguous. Of further interest to the geochemist are recent suggestions that these grains may have been present during the accumulation of planetary material and, hence, may be present in comets or even the more primitive chondrites. Their entire history, from their presumed origin in stellar atmospheres, through residence in the interstellar medium to a potential role in stellar or planetary formation, is just emerging as a topic for experimental and theoretical consideration.

Mineralogy and Texture

The major advance in the past 10 years has been the widespread application of the electron microprobe to the determination of mineral composition. This has contributed to the definition of the unequilibrated ordinary chondrites, which in turn led to the development of the concept of metamorphic effects in meteorites. Another area in which the probe has been extensively applied is in the determination of nickel contents in metal grains. These analytical data, together with experimental data on diffusion rates, permit estimates of cooling rates. Complete analysis of individual minerals has also meant that geothermometry studies on meteorites are possible.

Mineralogy New minerals are discovered in meteorites at an ever increasing rate. Admittedly, most of these are present in only trace amounts, but each can reveal something about the origin or evolution of the meteorites. Unfortunately, little effort has been expended to determine experimentally the conditions under which the minerals are stable. Such information obviously has a restricted application, but frequently the effort is worthwhile.

The distribution of minor and trace elements among meteorite phases will probably be of ever increasing importance. Experimental, theoretical, and analytical approaches are all required to determine those areas where the greatest return can be expected. Unfortunately, the available data are fragmentary, making projections difficult. A potentially useful thermometer is the distribution of iron and magnesium between clino- and ortho-pyroxene.

Chondrules The origin of chondrules remains an unresolved mystery in meteoritics. Most scientists now agree that they are frozen liquid droplets, but here agreement ends. The major question to be resolved is whether they are of "primary origin," the direct result of condensation in the nebula, or of "secondary origin," produced by some event such as lightning discharge, brief high pressure–temperature pulses, collisions among dust grains, or impact on the surfaces of protoplanets. This whole question is still sufficiently vague that all new approaches and insights must be seriously considered.

Metallurgy Iron meteorites usually contain a characteristic two-phase structure called a Widmanstätten pattern. Over the past 10 years, these meteorites have been studied in considerable detail. The observations, coupled with experimental work on nickel diffusion in iron–nickel alloys, have led to a major breakthrough, using the distribution of nickel across the two-phase structure as a means of determining the rate at which the meteorites cooled. Subtle distinctions in the patterns are correlated with cooling rates, which range between 0.1 and 100 degrees per million years. Additional studies of the effect of trace components on the estimated cooling rates are required to improve the accuracy of these estimates.

Paleomagnetism The study of paleomagnetism in meteorites has had an uneven history. Interest has been growing rapidly since the discovery that the more primitive meteorites may contain a record of the magnetic fields present in the nebula. These studies show considerable promise of revealing critical information on the strength of magnetic fields during stellar formation.

Tektites These peculiar, silicon dioxide rich, glassy objects, which are distributed worldwide, are frequently considered to be a type of meteorite, at least in the sense that they may be extraterrestrial. Current evidence, however, strongly favors the idea that they are terrestrial in origin, produced during terrestrial impacts of cosmic proportions—sufficient to remove a portion of the atmosphere. The mechanics of such impacts, as well as their frequency (1 per 15 m.y.) are problems of basic significance.

CHEMISTRY OF THE SOLAR SYSTEM

The Sun

The sun represents 98 percent of the mass of the solar system; its composition is thus, to a first approximation, the composition of the entire system. By comparing the composition of other samples with solar abundances, the geochemist determines the degree, direction, and the process of fractionation. Primitive Type I carbonaceous chondrites contain the relatively nonvolatile elements in proportions similar to those observed in the sun. One of the primary goals of geochemists is to determine and explain the differences between the sun, various types of meteorites, and the planets.

Information concerning the sun's composition comes from spectroscopic studies conducted mainly by astronomers, and solar-wind studies, conducted mainly by geochemists. Solar wind, in simple terms, consists of ionized atoms that are continuously emitted from the sun; during flares the amounts and energy of the solar wind increase measurably. All bodies within the path of the flare are potential collectors of solar wind. Rocks on the lunar surface have been exposed to this material for 3 or 4 billion years. Virtually their entire inventory of noble gases and much of their carbon appears to have been derived from solar wind. Some meteorites contain a mixture of fine-grained material enriched in solar-wind-derived noble gases, implying previous exposure. The tracks observed in the mineral grains of some meteorites indicate that the intensity and energy of the solar wind was higher in the past than it is today. However, it could not have been much more energetic without initiating nuclear reactions, the evidence for which would be revealed in studies of isotopic ratios. Extensive searches for evidence of proton-induced reactions in planetary material have so far been negative.

Comparative Planetology

The objective of comparative planetology is to develop unified theories concerning the evolution of the bodies in the solar system as well as the system itself. These studies draw on data obtained from all the planets, their satellites, the meteorites, the sun, and asteroids. Because of the paucity of data concerning most of these bodies, the field is still in its infancy.

Major progress has been in the development of theories of the evolution of planetary atmospheres. Comparison of the atmospheres of Earth and Venus reveals some interesting coincidences which, although poorly understood, are intriguing. The atmospheric pressure of Venus is about 100 times that of Earth; about 99 percent of its atmosphere is carbon dioxide, and most of the remaining 1 percent is nitrogen. If the earth's surface were as hot as that of Venus (about 500 °C), most of the sedimentary limestones would release carbon dioxide, yielding an atmosphere with roughly a 99/1 ratio of carbon dioxide to nitrogen. Thus, carbon dioxide and nitrogen are present in roughly the same amounts on the surfaces of both planets, but in one case carbon dioxide is combined with calcium and oxygen as limestone and in the other it is present as gaseous carbon dioxide. The Venusian atmosphere contains little water compared with what would be present in the terrestrial atmosphere if the surface temperature were 500 °C, sufficient to vaporize the rivers and oceans. A clearer understanding of these similarities and differences might be of tremendous importance. For example, it has been suggested that the high carbon dioxide content in the Venusian atmosphere is caused by a "runaway greenhouse" effect. The question arises: Is this condition unique to Venus, perhaps owing to its lack of oceans, or could it also happen on Earth? It appears possible that if the carbon dioxide content in the terrestrial atmosphere exceeds a critical level (for whatever reason), Earth's surface temperature would begin rising, which in turn would release additional carbon dioxide from the oceans and produce a self-generating thermal increase with disastrous consequences.

Venusian atmosphere also contains modest amounts of hydrofluoric and hydrochloric acids, rather unexpected atmospheric compounds, which has led to interest in the dominant gas–solid chemical reactions that might take place between the surface rocks and atmosphere. For these reasons, it would be of interest to know the mineralogy and composition of the soil and bedrock of Venus.

The atmospheric pressure on Mars is only about 1/200 that of Earth. The amount of water present is of interest, because Mars is probably the only planet with even a remote chance of supporting life. Recent high resolution photography of the Martian surface has revealed meandering valleys with tributary systems strongly suggesting the effects of flowing water. Current knowledge of the pressure–temperature conditions on the surface seems to rule out the possibility that liquid water could exist now. It has been suggested that every 25,000 years the polar caps on Mars, presumably solid carbon dioxide, vaporize and generate a modest atmosphere sufficient to allow liquid water to become stable. Geochemists and astronomers have also attempted to explain the red color of Mars as being caused by the interaction of water with iron in the crustal rocks and soil to produce a "rust," hematite (Fe_2O_3) or goethite (FeOOH).

With the exception of a relatively small research effort in astronomy, which involves Earth-based infrared analyses of planetary surfaces and atmospheres, the nature of future geochemical research on the inner planets will depend primarily on the space program and the type of missions chosen. Unmanned landing of instruments to perform remote analyses on the planetary surfaces will require the support of relatively large groups of instrumentation-oriented workers. On the other hand, the return of samples to terrestrial laboratories, through either manned or unmanned missions, will allow many workers to study the samples in depth and with much greater precision than that attainable using remote instrumentation. Several considerations favor unmanned sample-return missions over the other possibilities:

- The Apollo program has developed small-sample laboratory capability.
- The fine soil fraction often represents a broad sampling horizon, at least on the moon.
- Unmanned missions cost less.
- Unmanned missions could sample areas, such as polar regions, that might be hazardous and prohibitive for manned landings.
- Manned missions will be limited to Mars and the asteroids in the

near future; unmanned missions can probably sample Mercury, Venus, Mars, and the asteroids.

Comets and Asteroids   Current knowledge of the chemical composition of these bodies is superficial, although many contend that we already have samples of asteroids and comets in meteorites and "cosmic" dust. Reflectance spectroscopy of several asteroids has shown that they have compositions similar to both basalt and chondrites. With the possibility of future space missions being flown to cometary and asteroidal bodies, analyses of interest to the geochemist might be made of asteroid surfaces using passive X-ray fluorescence. Mass spectrometers capable of sampling the dust in comet tails might also be flown.

The Moon   The Apollo program has had a profound effect on geochemists, on their assumptions, on their concepts, on their activity, on their instruments, and on their future.

The single most important result of the geochemical exploration of the moon has been the foreshortening of the time scale of the accretion and initial differentiation of planetary-sized objects to, at most, a few hundred million years. For most investigators the biggest surprise of the sample-analysis program was the great age of crystallization of the rocks at the surface of the moon. It appears, through the study of strontium and rubidium isotopes, that separation into different rock types of the surficial material of the moon occurred shortly after, or perhaps even during, the time of formation of the moon itself.

Geochemical studies point to an igneous evolution of the moon that is quite distinct compared to the more commonly studied processes on Earth. In contrast to terrestrial igneous rocks, which exhibit a broad spectrum of chemical compositions, lunar rocks are more restricted in composition and fall into three categories: anorthositic gabbro, basalt, and K R E E P (an acronym for a rock, unusual relative to terrestrial species, rich in potassium, rare-earth elements and phosphorus). The anorthositic gabbro is the dominant rock type in the highlands, whereas the mare basins are filled with basalt. Fragments of K R E E P appear in the soils; the amounts seem to increase in and around Mare Imbrium, in the northwest quadrant of the moon, and drop off with increasing geographic distance from this region. Fragments of other rock types, including granite-like material, occur sporadically in the soils and breccias. Other rock types might easily turn up in the unstudied 90 percent of the returned samples.

All the rocks studied so far were formed in an environment deficient in oxygen and water. They are also deficient in other volatiles, such as lead, bismuth, indium, and carbon, compared to Earth and the meteorites, suggesting a moon-wide loss of volatiles. Noble metals, such as gold and palladium, also are conspicuously deficient in moon rocks compared to the content of terrestrial rocks. This may be related to the more reducing conditions of formation or to deficiencies of these metals in the moon as a whole. The large amounts of calcium, aluminum, titanium, and rare earths in the surficial rocks indicate enrichment at their source inside the moon, which has led to a consensus that the moon is relatively enriched in these refractory elements.

The study of lunar samples has had an impact on other research areas. The discovery of chondrule-like spherules on the lunar surface has lent support to the theory of the secondary origin of chondrules. Compositional variations of the rocks studied appear to support a model of heterogeneous accumulation, wherein the primordial material from which the planets were created is considered to have varied chemically in space and time, so that planets acquired chemical heterogeneity as they accumulated. The short time interval of accumulation and petrogenesis of the moon also places severe constraints on the thermodynamic conditions of accretion for the moon and, by implication, for the terrestrial planets as well.

Several ancient questions regarding the moon seem to have been solved, or at least are nearing solution, through the study of lunar samples. There appears to have been no life and very little water on the moon; impact cratering was obviously the dominant process in the modification of the lunar landscape. The only direct evidence of recent volcanic activity is reported gaseous eruptions from a single crater. Evidently the moon's igneous history stopped almost completely 3 billion years ago.

The age and composition of lunar rocks have placed severe constraints on two favorite theories of lunar origin: that of a daughter spun off Earth and of a stranger captured by Earth. The unearthly elemental abundances in lunar rocks speak against the concept of deriving the moon from the earth by fission. The great age of the moon rocks and the moon itself indicates that if it were captured, the event occurred early in the history of the solar system, at the time Earth and the moon were forming. This theory has, in a sense, merged with the alternative: Earth and the moon formed simultaneously as planet and satellite. There is no doubt that, when the voluminous data already available have been digested, relatively detailed and quantitative models of lunar formation will be developed. The termination of the Apollo manned-landing program is a suitable juncture at which to assess the present opportunities and future directions of lunar research.

Present Opportunities

Sample Analysis The program of sample analysis, started in the late 1960's during the Apollo program, should continue because only about 10 percent of the samples returned have been given more than a cursory examination. Work on the following specific areas of interest should be considered:

- Continued methodical examination of all samples for purposes of general classification and selection of unusual specimens for additional study.
- A search for phenocrystic materials (specimens with coarser crystals in a finer matrix) that may have formed at depth and, therefore, bear a more direct relation to the early fractionation processes.
- Continued study of small fragments in the soils and in breccias. This is significant because these small fragments represent a broader areal sampling than larger specimens. This aspect of the future lunar-sample analysis program will emphasize, even more than before, the importance of microanalytical techniques. It will rest heavily on the development of instruments such as the ion-source mass spectrometer, capable of performing precise trace element and isotopic analyses directly on solid specimens.

Data Reduction and Interpretation The Apollo program, in addition to producing a mass of laboratory analytical data, has also produced a wealth of analyses conducted by means of remote instruments flown in orbit around the moon. Many of these orbital analytical data should be studied to ascertain their relation to variations in rock types and to lunar topographical and geophysical features.

The rapid influx of laboratory analytical data has left researchers little time for the correlation of their data with that of other investigators. For example, the correlation of carbon, sulfur, nitrogen, and lead isotopic ratios and the abundance of mercury and other volatiles in fine soils as indicators of vapor fractionation are of interest; but because these analyses are generally performed by different laboratory groups, there has been little comparison and discussion of these data. The data bank of lunar-sample analyses is in a form that can be handled by computers and should be maintained and built up. Multivariate statistical analyses of these data and their correlation with the analyses produced remotely from lunar-orbiting vehicles should be fruitful.

Future Directions

With the increased number of lunar-sample analyses, especially complete analyses of fine soil particles, together with orbital analytical and geophysical data, a more complete knowledge of the origin of the lunar rocks and the moon itself is certain to be built up. This knowledge should stimulate more penetrating questions to be answered on future missions. It is possible that sophisticated unmanned devices to perform lunar analyses with adequate precision can be further developed. Small soil samples can be returned by unmanned space missions. This situation is similar to that envisioned in the planetary exploration program and, in this case, would permit the exploration of the far side of the moon as well as its polar regions.

Extraterrestrial Organic Matter

Interstellar Extraterrestrial organic matter occurs in the interstellar medium, in the atmospheres of planets, on the surface of the moon and in meteorites. Thus far, only the moon and meteorites have been sampled for analysis. Radio astronomy, especially during the last few years, has provided evidence for at least 20 different kinds of simple carbon-containing compounds in various regions of the interstellar medium. The discovery of such chemical species as hydrogen cyanide, cyanoacetylene, and formaldehyde suggests that even more complicated molecules synthesized from these reactive precursors may be present. Although conditions in the interstellar medium are difficult, if not impossible, to simulate, an effort could be made to create actual or theoretical analogues of the interstellar medium in an attempt to understand the processes and to predict what chemical species may be present.

Planetary The principal constituents of the atmospheres of Venus, Mars, and Jupiter have been determined, but the trace constituents are still to be studied. Methane, the simplest of the possible organic compounds, does occur in the atmospheres of Jupiter and Mars. The geochemist can contribute to understanding planetary atmospheric processes by attempting to simulate the organic chemistry that takes place within the atmospheres of the various planets. Chemical exploration of planetary atmospheres by means of spectroscopy and by probes performing direct analyses will become more active in the future and requires direction from the geochemist. These studies should provide data with which to improve models of the origin and evolution of planetary atmospheres.

Lunar Material returned from six lunar missions has provided an unprecedented opportunity to examine pristine extraterrestrial samples. Examinations have shown that organic matter is present in extremely low concentrations. The simple organic gases, methane and ethane, are present, but the occurrence of more complex organic molecules is questionable. Recent work has shown that indigenous hydrogen cyanide is also a constituent, possibly derived from past cometary impact. Given the proper environment, this molecule could be the precursor of more complex substances such as amino acids. Its role in lunar carbon chemistry should be ascertained. The contributions of solar wind and meteorites to lunar carbon chemistry appear to be important areas of continued lunar research. The moon is an excellent natural laboratory for the examination of solar-wind phenomena.

Meteoritic Carbonaceous chondrites contain significant concentrations of organic compounds, but these meteorites are readily contaminated with biospheric organic materials that apparently invade the specimens almost immediately on contact with the earth. Recent work has shown that in some instances biospheric contamination has been minimal. In these instances the composition of the extraterrestrial organic material may be accurately determined.

Future work should concentrate on determining the significance of meteorite organic materials to extraterrestrial processes. The mechanism of origin of individual compounds, the physical and chemical conditions of origin, the metamorphic history of the meteorite body, and the relationship to the origin of the solar system all constitute valuable directions for research on the organic chemistry of meteorites. Laboratory synthesis under simulated conditions has proved fruitful in recent years. Organic compounds formed by catalytic reactions using carbon monoxide and hydrogen as reactants bear a remarkable resemblance to the compounds observed in meteorites. Carbon isotopic studies of both observed and synthesized compounds should shed new light on the conditions and mechanisms of formation.

SELECTED READINGS IN EXTRATERRESTRIAL GEOCHEMISTRY

Anders, Edward. Meteorites and the early solar system. Ann. Rev. of Astron. and Astrophys. 9:1–34, 1971.
Hinners, N. W. The new moon: A view. Rev. of Geophys. 9:447–522, 1971.
Levinson, A. A., and S. R. Taylor. Moon rocks and minerals. Oxford: Pergamon Press, 1971.
Mason, Brian. Meteorites. New York: John Wiley and Sons, 1962. 274 p.
Mason, Brian, and W. G. Melson. Lunar rocks. New York: John Wiley and Sons, 1970.
Wood, John A. Meteorites and the origin of planets. New York: McGraw-Hill, 1968.

Solid Earth Geochemistry

PURPOSE

The purpose of solid earth geochemistry is to determine the chemical composition and distribution of materials within the earth and to understand the processes by which the composition and distribution of materials within the earth change in space and time.

THE EARTH'S INTERIOR

On the basis of geophysical properties, the earth can be divided into three concentric units: core, mantle, and crust. Dimensions, masses, and mean densities of these units are listed in Table 1. Only the crust, making up about 0.3 percent of the total mass of the earth, can be extensively sampled and chemically analyzed.

Our knowledge of the composition of the remaining 99.7 percent of the earth is almost entirely indirect and is based primarily on geophysical observations of the earth combined with measurement of the physical properties of rocks and minerals in the laboratory and on the compositions of meteorites, the moon, and the sun.

The upper part of the continental crust is granodioritic in composition, richer in radioactive elements (uranium, thorium, potassium),

TABLE 1 Geophysical Properties of the Earth–Dimensions, Masses, and Mean Densities

Unit	Thickness or Radius (km)	Mass (g)	Mean Density (g/cm³)
Oceans	4	1.4×10^{24}	1.0
Oceanic crust	7	7.0×10^{24}	2.8
Continental crust	40	1.6×10^{25}	2.8
Mantle	2870	4.1×10^{27}	4.6
Core	3480	1.9×10^{27}	10.6
Earth (total)	6371	6.0×10^{27}	5.5

silica, and alkalies than the upper mantle, the oceanic crust, and the lower part of the continental crust. The mantle is ultramafic in composition, and consists essentially of magnesium- and iron silicates, with relatively low concentrations of such elements as calcium, aluminum, potassium, sodium, and uranium. The oceanic crust is mafic in composition, intermediate between that of the continental crust and that of the mantle.

BASIC PROBLEMS

In recent years the concept of plate tectonics has developed; that is, that the crust and upper part of the earth's mantle are made up of a small number of large rigid plates that move relative to each other. During this movement, material from the mantle rises along great oceanic fractures, causing oceanic ridges. As the material continues to rise and solidify, the plates on either side of the ridge move laterally away from the ridge, causing separation of one plate from another (seafloor spreading). Where they come together, one plate moves downward steeply, and the other overrides it. This model of plate behavior has provided a framework for relating many, but by no means all, crystalline rock-forming phenomena to a globe-encompassing cycle in which large amounts of material are transferred from the mantle to the crust, modified in various ways by near-surface processes and then returned to the mantle. In this context, the following are some of the long-standing problems in solid earth geochemistry:

- *The origin of petrogenic provinces.* Petrogenic provinces are worldwide occurrences of characteristic suites of the three main rock types—igneous, metamorphic, and sedimentary—that are closely related in space and time. Many of these suites can now be related to specific environments and processes that occur within the plate-tectonic cycle.

- *The origin of igneous rocks* (rocks crystallized from silicate melts). Unanswered questions related to this topic include: What is the bulk chemical nature of the source materials within the mantle or crust; what are the conditions of temperature, confining pressure and abundance of volatiles that generate molten material at depth; and what are the processes that have modified the composition of such melts during transport and eventual emplacement under changing conditions of temperature, pressure, and chemical environment?

- *The origin of metamorphic rocks.* Unanswered questions related to rocks recrystallized in the solid state include: What was the source material, and what was the physical and chemical environment under which recrystallization took place; to what extent have the original compositions been modified by additions or subtractions of material; and how much chemical interaction is there between different levels in the earth's crust and between the crust and mantle during metamorphism?

- *The nature and origin of ore deposits.* Ore deposits are rocks that have unusual and economically significant concentrations of certain elements. An understanding of the processes by which these unusual concentrations formed should make it possible to recognize areas where geologic conditions are especially favorable for ore deposits, and thus provide targets for exploration.

- *Nature of the mantle.* Unanswered questions related to this topic include: Are there systematic differences in composition of the upper mantle underlying stable continental crust and that underlying oceanic crust; in what amounts and in what forms are water and carbon dioxide present in the mantle; and what is the distribution of radioactive elements such as uranium, thorium, and potassium within the mantle?

METHODS AND TECHNIQUES

Thermodynamic Calculation

Knowledge of the compositions and the thermodynamic properties of phases in a system at elevated temperature and pressure allow calculation of self-consistent equilibrium assemblages as functions of whichever intensive variables may be assumed to have been imposed by the environment. These variables include temperature and confining pressure but may also include partial pressures (or concentrations) of volatiles in a fluid phase and chemical potentials of mobile elements in the solid or crystalline phases. Comparison of actual mineral assemblages observed in a suite of rocks with the calculated assemblages may set limits on the temperature and confining pressure, and on the gradients in those variables during formation of the rocks. It may also be possible to identify rocks in which the composition of an intergranular fluid phase or the chemical potentials of relatively mobile components have been controlled by the local mineral assemblage versus rocks in which the environment has controlled these variables, or in which certain components have been redistributed in response to local chemical potential gradients. The use of computer processing will become more and more important for calculating and predicting complex equilibrium relationships. Also, studies of solid-solution (multiphase equilibrium) theory should be continued for the guidance they provide to the experimentalist.

Although recent efforts have been made to consolidate, fill in gaps, and eliminate inconsistencies in thermodynamic data for the rock-forming minerals and volatiles at high temperature and pressure, more work needs to be done. More precise calorimetric data (heat capacities, entropies, and heats of solution) are still needed for a number of common rock-forming minerals, particularly those showing more or less complex nonstoichiometric or solid-solution behavior (for example, amphiboles, micas, and garnets). Well-characterized synthetic materials of end-members and intermediate compositions are either available, or could probably be obtained, in sufficient quantity for calorimetric determinations.

Phase Equilibrium Studies

Equilibrium–reaction boundaries between phases determined in the laboratory provide calibration points for the interpretation of phase assemblages in natural rocks in terms of temperature and pressure of formation. The pressure and temperature capabilities of several types of experimental apparatus are indicated in Figure 2. Because the pressures at the continental crust–mantle boundary and the core-mantle boundary are about 10 kbar and 1400 kbar, respectively, it is clear that conditions in the crust and upper mantle can be investigated by several experimental techniques, but in the lower mantle and core (region below 500-km depth) they can be investigated only by shock-wave experiments of about one millionth of a second duration. Development of static ultrahigh-pressure equipment will be pivotal in furthering our understanding of the physical and chemical properties of the core and lower mantle.

Recent high temperature–pressure experiments, in which the electron microprobe has been used to analyze the phases produced in the experiments, have been particularly helpful in deciphering melting relationships in compositionally complex systems. In fact, the electron microprobe, with its ability to analyze quantitatively for major elements on a micron-size scale, has proved indispensable in understanding solid-solution series in multicomponent systems. The compositions of pyroxene solid-solution phases that have been determined in this way appear to offer a promising basis for estimating the temp-

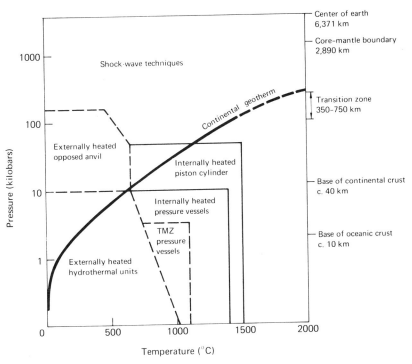

FIGURE 2 Capabilities of high pressure–high temperature apparatus. A kilobar is roughly 1000 atmospheres. Heavy line represents geotherm underlying stable continental region. (TMZ = titanium–molybdenum–zirconium alloy).

erature and pressure of formation of mantle-derived rocks. Other geothermometers and geobarometers have been suggested and will undoubtedly be developed, based on the complex solid-solution behavior of other mineral phases.

The pressure–temperature conditions for the dehydration reactions, partial melting, and solid-state phase transitions that provide a framework for the facies classification of metamorphic rocks (subdivision into groups based on presence of characteristic minerals) are sufficiently well determined in many instances. Although a great deal of detailed work in hydrothermal systems (those containing water, carbon dioxide, and other essential constituents) remains to be done, much of the work is time-consuming and needs continuing appraisal of expected returns in contrast to required input.

Interest in hydrothermal experiments has recently centered on mixed-volatile reactions in which the composition of the fluid phase is a variable. Experiments with carbon dioxide–water fluids are essential for understanding the metamorphism of carbonate-bearing sediments, and experiments with hydrous-salt fluids will provide a basis for understanding the role of the intergranular fluid as a transporting and catalyzing medium for compositional and textural changes in solid rocks. Techniques for hydrothermal experimentation under controlled or buffered oxygen pressure have been available for some time, and more recently, buffering techniques have been developed for chloride, fluoride, and hydrogen-ion activities. It may be that specific ion-membrane electrodes, similar to those developed for oxygen, can be utilized in laboratory studies of natural rocks to study the partial pressures of chlorine, fluorine, sulfur, and other elements in various natural environments.

Kinetic Studies and Calculations

Our understanding of the equilibrium properties of earth material systems is in a much more satisfactory state than our understanding of the processes by which the equilibrium state is approached or disturbed. Kinetic or rate studies of rock-forming processes are in their infancy, and there is certain to be a great increase in experimental determinations of rates of mass transfer, atomic diffusion, crystal growth, ordering, and disordering. Prediction of rates from first principles is more difficult, because it requires knowledge of properties of the system and reaction mechanisms at a molecular level. A start will be made in this direction when data on bonding energies, site distributions, and defect concentrations in mineral phases become available; but our lack of empirical data on the molecular properties of silicate-melt phases and crystal surfaces at elevated temperature and pressure guarantees that this will be an area of continuing and expanding research in the near future.

Stable Isotope Geochemistry

The natural variations of isotopic composition of a number of stable isotopes—oxygen, hydrogen, and carbon, for example—have proved useful in solid earth geochemical studies. Uses range from "fingerprinting" or characterization studies (in which, for example, the oxygen isotopic ratios of a metamorphic rock may be used to decide whether it was derived from an igneous or sedimentary parent), to studies of volatile exchange (in which, for example, the distinctive oxygen and hydrogen isotopic ratios of groundwaters can be used to evaluate the amount of exchange between igneous rocks and groundwater reservoirs), to geothermometry studies (in which temperatures can be estimated by the partitioning of the oxygen isotopes between coexisting mineral phases). Silicate phases are rich in oxygen, and volatile phases are rich in hydrogen; consequently these isotope pairs are ideal for studying interactions between solids and volatiles. Because of the low concentration of carbon in many igneous and metamorphic phases, carbon isotopic studies have received comparatively little attention; however, carbon is potentially useful in studying the volatile-phase geochemistry of metamorphic rocks. Studies of carbon in igneous rocks would probably also be rewarding, although somewhat demanding technically. One could look, for example, at the quenched glass of submarine basalts (lavas erupted under water), which appears to trap a sample of the dissolved-gas load of the basalt melt, including gases containing elements such as hydrogen, fluorine, chlorine, sulfur, oxygen, and carbon. In addition to carbon, oxygen, and hydrogen isotopic measurements, sulfur isotopic ratios can also be utilized in this way.

Trace Element Studies

A variety of sensitive and reliable techniques now exist for quantitative analysis of many of the trace (≤ 0.1 percent) elements found in natural materials. The trend recently has been toward improving the techniques to work with smaller sample sizes (milligram samples are sufficient in many instances) and toward chemical schemes that analyze many related elements simultaneously. Pure samples still cannot always be obtained in sufficient quantity for analysis. For example, laboratory measurements of trace-elements partitioning between minerals are greatly hindered at present by the small amounts of material produced experimentally and the extreme difficulty of separating fine-grained phases. In these instances, an *in situ* analysis technique becomes important. The ion probe may eventually be a means of performing this type of analysis, because it has the potential of analyzing for elements at concentrations of parts per million on a micron scale. Fission-track mapping has proved useful, at least with one element (uranium), for mapping low-concentration levels with spatial resolution of a few microns.

Often the major problem in trace element studies is to obtain sufficient numbers of samples that have been well characterized from a major-element point of view. It is gradually becoming apparent that the combined use of major and trace elements results in a much more powerful approach than the use of either one separately. This approach frequently creates complications because there are few laboratories with facilities for using both methods. As a result, many effective cooperative programs have been initiated between laboratories, and more should be encouraged.

Geochronology

Time is, of course, a key parameter in studying the geochemical evolution of the earth. Ages can be determined by use of various natural radioactive parent–daughter systems. In general, igneous events (crystallization of silicate melts) can be dated with more confidence than metamorphic events. For example, young lavas can be dated by the potassium/argon method. Intrusive igneous rocks that have not undergone later metamorphism can be reliably dated by the potassium/argon or rubidium/strontium method on minerals such as hornblende and muscovite. Older intrusive rocks, even those that have been through a later metamorphism, usually give reliable ages by whole-rock rubidium/strontium or zircon (the mineral $ZrSiO_4$) uranium/lead techniques. Precision and accuracy of a few percent can frequently be achieved. The zircon uranium/lead technique has recently shown marked advances because of improvements in analytical technique; ages as young as a few million years can be determined, and time resolution of less than a million years can be attained even for the oldest terrestrial rocks. Problems can, however, occasionally show up in any of these techniques, and methodology needs more refinement in some cases. For example, submarine basalts are difficult to date owing to alteration effects and excess argon incorporation; whole-rock rubidium/strontium ages are sometimes disturbed by subtle effects that are not well understood; zircon ages may be in error if a relict zircon component is inherited by contamination from older rocks. Also, particular rocks pose particular problems; older mafic rocks, for example, are difficult to date by existing techniques. Whole-rock dating with thorium/lead may be a feasible method to develop for mafic rocks. There is also some indication that mafic rocks have lutetium/hafnium ratios that would favor development of the lutetium/hafnium method for such rocks, though analytical methods are not well established for hafnium.

Although the use of radiometric techniques is fairly well developed for dating of simple igneous events, age determination is a much more complex problem for metamorphic rocks. The difficulty lies in determining what a measured age actually means for metamorphic rocks. For example, potassium/argon or fission-track ages of minerals such as biotite are sensitive to temperature, owing to low-temperature fission-track annealing and argon loss by diffusion. Thus a potassium/argon age of a metamorphic mineral generally is a "cooling age," or the time at which the mineral (and surrounding terrain) cooled below a temperature where argon diffusion was no longer significant.

Various minerals may "shut off" at different temperatures, thus allowing a time–temperature history to be developed for a metamorphic area. Most minerals, however, lose argon and strontium at temperatures well below peak metamorphic temperatures, and so will record only the waning stages of metamorphism and not the initial or peak stages. Certain mineral-dating methods, for example uranium/lead in zircon, show a wide variation in temperature response, depending on variables such as radiation damage and composition. The whole-rock rubidium/strontium dating method has been shown to be resistant to metamorphism in certain cases. For example, rocks that initially crystallized at high temperatures (igneous rocks and high-grade metamorphic rocks) may retain their whole-rock rubidium/strontium age through a later lower or similar grade metamorphism. Under certain conditions, perhaps related to volatile content or degree of deformation, the whole-rock ages may be affected by a later event. Thus the event measured by a whole-rock rubidium/strontium age is not necessarily defined. For metamorphosed sedimentary rocks, the picture is even more complicated. Can these rocks yield a sedimentary, or even presedimentary age, and what conditions are required during metamorphism of a sedimentary rock to reset its age and provide a true time of metamorphism? General understanding of radiometric dating of metamorphic rock is fragmentary. Because of their importance in the evolution of tectonic terrains, it seems likely that much of the methodological development in radiometric dating in the next decade will be addressed to the problem of metamorphic rocks.

THE PLATE-TECTONIC CYCLE

The plate-tectonic model has suggested underlying relations between many of the basic problems of solid earth geochemistry, and the model has in turn been modified and constrained by recent investigations in solid earth geochemistry. Only a part of solid earth geochemistry is directly involved in the plate-tectonic cycle (or perhaps involvement is not yet established), but the cycle provides a geographic array of geotectonic environments useful in discussing the processes and problems of solid earth geochemistry. The plate-tectonic cycle is best understood in its present or recent style; one of the most important threads of future research will be to determine whether plate-tectonic cycles were present throughout earth history, and if so, what were their styles.

The basic features of the plate-tectonic cycle are shown in Figure 3. Geometrically, the model supposes a relatively rigid "plate" (lithosphere) of thickness up to several hundreds of kilometers, overlying a more yielding zone of partly molten material (asthenosphere). Molten or partly molten material rises from the asthenosphere along the zones of the great linear midoceanic ridges, where mafic silicate magma is separated from the mantle and provides a capping or crust. This newly formed oceanic crust and underlying mantle both move away from the zone of upwelling and slowly cool to form the rigid plate. This part of the process has been termed "seafloor spreading."

At the opposite edge of this plate is often a zone of underthrusting (subduction) where the plate, with its capping of mafic igneous crust and accumulated sediment, returns to the asthenosphere. This zone is marked by the great ocean trenches and their bordering volcanic arcs; the underthrust plate is delineated by steeply inclined zones of concentrated seismic activity (Benioff zones). Material balance is presumably maintained during plate subduction by a counterflow of asthenosphere back toward the upwelling zones under the ridges. This process may be called the plate-tectonic cycle. The driving forces for this cycle are probably some form of thermal convection; however, the kinematics of the sublithosphere motions are at present poorly understood. It is as yet unknown, for example, whether the material of the subducted plate is ever actually recycled back to the ridges, or whether it is permanently removed from the cycle, thereby dooming the cycle to eventual exhaustion.

The plate-tectonic model thus provides several distinctly different tectonic environments (petrogenic provinces) for the formation of characteristic suites of igneous, metamorphic, and sedimentary rocks, and one important task of solid earth geochemistry is to describe or characterize the rocks from each of these provinces, evaluating their similarities and differences. Once established, this characterization will be the basis for study of processes of formation and modification. The timing of various events in the cycle is important both in a descriptive sense and for understanding of the formation processes. Finally, the rock assemblages from the older geologic record can be interpreted in light of the characterizations developed for the recent tectonic cycle.

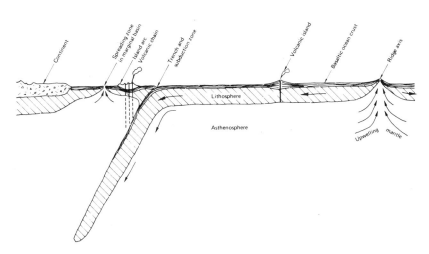

FIGURE 3   Basic features of the plate-tectonic cycle.

though much of the data (and speculation) bearing on the geochemistry of the mantle and rocks at the base of the continental crust are based on analyses of inclusions from kimberlite pipes, the processes that produce kimberlite and carbonatite magmas in the mantle remain obscure.

Continental Margins

Present margins of continents are of two types: they may be active (such as the west coast of South America), marked by a trench with a fringing volcanic chain or arc, or they may be inactive (like the east coast of North America), in which great thicknesses of nearly flat-lying limestones and shales may be deposited on the continental shelf. When inactive continental margins become active, the continental shelf sediments may be greatly deformed, but, because of the relatively light continental crust that serves as their basement, they will tend not to be carried down the subduction zone. The sedimentary section may, however, undergo tectonic thickening, with burial and metamorphism of structurally deep parts of the section at high temperature and high pressure followed by rapid erosion and unloading to expose the newly formed metamorphic rocks at the surface within a relatively short time span.

Arc-Trench (Subduction) Province

Most of the development of continental crust, with its wide range of pressure–temperature environments, probably occurs in this complex province. A range of sediment types is present, including those deposited locally on the earlier inactive continental shelf margin, those deposited in the trench environment, and those carried in on the subducting plate. Metamorphic environments range from those associated with the surface of the downgoing plate (relatively low temperature, but high pressure) to those closer to the center of the zone of active plutonism and volcanism (high temperature, low pressure). The igneous rocks exhibit a wide spectrum of compositions, apparently related to the geographic position in the arc relative to the underlying seismic zone. (The causal nature of this relationship is obscure at present.) Many hydrothermal ore deposits also show geographic zoning, and their sources may in some cases be traced back to the basal metalliferous sediments formed on the spreading ridges.

Characterization of Rock Types

The general chemical and mineralogic nature of rocks from the various provinces is fairly well established. In many cases, the rocks of a given province are distinctive; thus the calc-alkaline volcanic suite is found in island-arc provinces but not in spreading-ridge or oceanic-island provinces. In other cases, however, rocks of similar bulk composition occur in all provinces; that is, tholeiitic basalts. The trace element patterns of these basalts are frequently diagnostic of the province of origin. The large-ion-lithophile (LIL) trace elements—such as rubidium, barium, strontium, uranium, and rare earths—have received the most attention for such characterization of volcanic rocks, partly because highly precise and sensitive analytical techniques are available. Interest is growing also in the transition-metal elements, as they are concentrated in crystalline phases relative to melts, in contrast to the LIL elements. Potentially, any element or group of elements could be useful for characterization purposes.

Rock types for which diagnostic chemical data are badly needed are the alkali basalts and the ultramafic and metavolcanic rocks of ophiolite complexes. On a finer scale, since rocks within a given province may also show geographic variations, a much larger body of data is needed to outline these variations. Especially critical is the need for trace element data on the intermediate to siliceous volcanic and plutonic rocks, and both major and trace element data on the metamorphosed sedimentary and volcanic rocks of island arcs.

Igneous Processes

The major-element character of magmas is controlled in the zone of melting by the mineralogy of the mantle source rock, which is in turn governed by the pressure–temperature conditions and the bulk composition of the material. A large body of data exists on the equilibrium relationships of the common silicate minerals at pressures and temperatures corresponding to zones of partial melting in the mantle. The general correspondence of tholeiitic basalt with shallow melting zones (such as on oceanic ridges and in island arcs over shallow parts of the subducted plate) and more alkaline basalt with deeper melting zones (such as under oceanic islands, and in island arcs over deeper parts of the subducted plate) is in good accord with predictions from phase-equilibrium data. Thus in a general way, magma composition may be used to infer mantle composition and the pressure–temperature conditions of melting.

Trace elements and isotopic patterns are important in characterizing the sources of magmas because trace elements are not partitioned under stoichiometric constraint; they are effectively decoupled from major-element behavior and in many cases provide information that cannot be derived from the major elements. For example, one might conclude from the worldwide similarity in composition of various basalt groups that the mantle sources were also similar in composition. However, no model for partial melting exists that can produce the observed large variation in trace element content of these basalts. It is thus fairly well established that the mantle is chemically heterogeneous. The variability of lead- and strontium isotope abundances in these rocks leads to the same conclusion (both strontium and lead have isotopes of variable abundance formed by decay of the radioactive parent rubidium and uranium-thorium isotopes, respectively). Furthermore, the lead isotopic data provide some estimate of how long this mantle heterogeneity has existed, and in most cases this time is of the order of one billion years or more. It is possible that the chemical and isotopic heterogeneities are related to previous extractions of magma or to additions of sedimentary or altered igneous material by subduction during earlier episodes in the plate-tectonic cycle.

As another example, the major-element composition of a basalt does not vary strongly as the degree of partial melting increases; that is, 10 percent melting and 20 percent melting may produce liquids of similar composition. However, these two liquids differ by a large amount (about a factor of two, in this instance) in their content of most of the LIL elements. More precise statements about the degree of melting involved in magma production, and of the particular mineral assemblages being melted, will depend on much more detailed knowledge of the effects of pressure, temperature, and composition on the partitioning of trace elements between various minerals and silicate melts. Because of the importance of this area of study, a concerted effort should be made in the immediate future to secure data of this type. In many instances, the measurements and interpretation may require development of more sophisticated analytical facilities, such as ion microprobes.

Magmas differentiate (undergo partial crystallization) during ascent, intrusion, storage in magma chambers, and eruption. The differentiation process is interesting both as a source of information about magma evolution in general and as an effect to be "removed" so that source chemistry can be studied. Where the nature of crystallizing minerals can be observed (or inferred), the major element chemistry of a series of lavas or of zoned plutons can be used to evaluate the differentiation process. Trace element patterns also provide constraints on this evaluation, and several studies have had notable success in deriving a differentiation scheme for lava series consistent with all the major and trace elements involved in the studies. Differentiation processes may also take place at depth, however, where crystallizing phases may not be preserved for observation. In these instances, trace elements provide particularly strong constraints on possible differentiation schemes; elements that concentrate in particular phases will be strongly depleted during removal of any of these phases from the liquid. Again, this approach is hampered by the lack of knowledge of trace element partition coefficients. Whereas some temperature effects have been documented, there is as yet little empirical or theoretical evidence for predicting the effect of pressure on partition coefficients.

Other processes, such as assimilation of materials encountered *en route*, wall–rock exchange, and volatile transfer, may affect magma composition. Assimilation or contamination can be modeled as a simple mixing process, and both major and trace elements are useful in constraining such a process. Radiogenic isotopes such as strontium and lead are particularly useful, because their composition in the magma will frequently be different from that in the country rock. Although simple models involving direct contamination by observable country rocks are sometimes sufficient to account for particular peculiarities of magma composition, the complexities of modeling are increased enormously when one considers the unknown compositions of deeper crustal contaminants and the possibility of selective contamination. How can possible deep-level contamination of magmas be identified if it is not known whether the wall rock is mafic or intermediate in composition, or whether it is normal or depleted with respect to the radioactive parent elements that control the isotopic ratios of the lead and strontium in the rock? Relatively nonradiogenic strontium is frequently used as evidence for a mantle source of particular magmatic rocks. How can a direct lower crustal origin be ruled out when the strontium composition of the lower crust is not known? Contamination models become even less tractable when it is considered that certain wall-rock elements may exchange with a magma to a larger degree than others—because of various differences such as migration rates, solubility in aqueous phases, and locations in particularly unreactive phases.

Many of these latter uncertainties are involved in the problem of volcanic-arc magma genesis. Are these magmas derived by melting of the downgoing lithospheric plate? Is contamination by sediment involved, and if so, is the contamination representative or selective? Or are the magmas primary melts of the mantle overlying the subducted plate, perhaps with additions of aqueous phases and dissolved elements from the subducted plate? Is there exchange or contamination of primary magmas with higher level mantle or with the deeper crustal sections of the arc? Although it is obviously important to be able to deal with these possibilities, it seems unlikely that the necessary level of sophistication can be achieved in the near future.

In studying igneous processes, complications also arise because of kinetic or mass-transport mechanisms. The crystallization paths of an igneous intrusive, for example, can be fairly well specified under conditions of melt–crystal equilibria; however, there is little information on the parameters that determine whether such equilibria exist. In some instances, minerals crystallize in complete equilibrium with the melt, whereas in others they may be zoned and exhibit only surface equilibrium or some intermediate state. It is not known to what extent the melt remains homogeneous; or whether diffusion gradients exist in the liquid near crystallizing minerals. What determines the extent to which liquid trapped between accumulating minerals remains in communication with free liquid? There is essentially no information on the nature of the diffusion process in natural silicate liquids and its dependence on various environmental parameters. Techniques in other disciplines could be directly applied to these problems in geochemistry, and this field needs investigation.

A related problem is the migration of hydrothermal fluids exsolved from magmas and perhaps of aqueous fluids derived from outside a cooling intrusive. Evidence from oxygen isotopic studies suggests that many intrusive igneous bodies may experience a convective flow-through of aqueous fluids from the wall rock, making it possible that two-way exchange of dissolved elements between intrusive and wall rock might be facilitated. Obviously, solubilities of elements in these fluids and rates of migration of both solvent and solute become important parameters in understanding changes like late-stage igneous alterations and hydrothermal ore derivations.

Another unknown area of mass-transport phenomena involves the accumulation and movement of magma from partial-melting zones in the mantle. Current models of basalt genesis that consider trace element constraints suggest that some basalt types are the result of small degrees of partial melting, perhaps as low as 1 percent. It is difficult to envision a process by which such melts, probably occurring first as films along grain boundaries, could coalesce into larger coherent volumes before transport to the surface. Such a process may occur, but there is as yet no adequate model to replace the present intuitive guesses. The nature of the actual magma-ascent process is unknown, but is of obvious importance because the opportunity for re-equilibration of melt with mantle host during ascent must depend on the geometry of the process (such as surface-to-volume ratios). Progress in this field could come from both theoretical and simulated laboratory models. Such studies are also important beyond the field of geochemistry, because the physical properties of the asthenosphere, such as seismic parameters and viscosity, probably depend critically on the nature of partial melting and magma-coalescence processes.

Metamorphic Processes

The character of a metamorphic rock depends on the composition of the original material and on the pressure–temperature conditions (and volatile content) of the environment during metamorphism. Equilibrium relationships are known for minerals of most of the common metamorphic assemblages, and to the extent that equilibrium is achieved in nature, the pressure–temperature conditions of metamorphism can be inferred. For example, two distinguishable metamorphic belts occur over subduction zones in some island arcs, and can be interpreted in terms of contiguous high pressure–low temperature and low pressure–high temperature environments. This pairing of pressure–temperature environments may be of use in determining directions of dip for the subduction zone. Additional information on temperature conditions can be derived from trace element and isotopic data. For example, partitioning of trace elements between given mineral phases may be temperature-dependent and, with proper calibration, could be used as a geothermometer. The partitioning of oxygen isotopes has been used successfully in this regard. More work is needed to develop pressure indicators. Here again, partitioning of certain trace elements may provide this information. The high pressure "blueschist" assemblages are missing from Precambrian rocks; does this mean that subduction or other tectonics capable of introducing rocks to high pressure were missing? Or have the blueschist assemblages been washed out of the Precambrian by high temperatures? An independent means of establishing pressure conditions would be a great asset in understanding Precambrian tectonic processes.

The development of metamorphic mineral assemblages obviously depends on how fast elements can rearrange themselves. Are metamorphic textures (grain configurations) really equilibrium features, or are they more likely to be caught in transit because of kinetic limitations? To what extent is metasomatism (long-distance chemical transport) capable of large-scale conversion of one rock type into another? Are migmatites (banded high-grade metamorphic rocks) formed by solid-state diffusion, segregation of local partial melts, or by injection of material from outside? To answer these questions requires detailed knowledge of diffusion coefficients under a variety of natural conditions; that is, as a function of pressure, temperature, fluid pressure, composition, and mechanical state of the rocks and minerals. Reliable data in this field are sparse, and it is expected that a major growth of emphasis in geochemistry in the next decade will occur in this general area.

It appears that the distances over which elements migrate in the solid state under relatively dry high-grade metamorphic conditions are limited. Certain mineral phases may have exceedingly small diffusion coefficients for their constitutional elements; for example, compositional zoning in garnet is common, indicating total movements of a centimeter or two. At lower metamorphic grades, adjacent rock units may show lack of oxygen- and strontium isotope equilibration, but at higher grades isotopic mixing tends to take place throughout larger and larger volumes of rock.

During progressive regional metamorphism of sediments, volatiles are released and are driven out of the rocks. The intergranular fluid, whether present as a phase or adsorbed surface film, acts not only as a transport medium for volatiles and materials complexed by them, but also as a catalyst for recrystallization, equilibration, and exchange of material between and within volumes of rock. Although the ratio of intergranular fluid to total rock at any given time may be vanishingly small, essentially the entire rock mass may have reacted with the intergranular fluid. A systematic study of fluid inclusions in metamorphic rocks may shed some light on these processes. Knowledge of the surface chemistry of volatile-rich surfaces at high temperature and pressure is almost nil, and basic data for the chemical and thermodynamic properties of volatile-rich fluids under these conditions are just beginning to be available. Future work on kinetics of metamorphic and late-stage igneous processes will depend on a great increase in this type of data.

Although it is at present difficult to specify the kind of migrations that can be expected in the presence of either a static or dynamic volatile phase, evidence from the radiogenic argon content of rocks suggests that an inert gas component is generally highly mobile; during metamorphism of old crystalline rocks, much of the accumulated radiogenic argon is released from the minerals and migrates out of the region, presumably along grain boundaries, cracks, and fissures. Under certain conditions, however, the radiogenic argon pressure in a metamorphic terrain may remain high enough to be retrapped into minerals during cooling of the terrain, producing easily observable anomalies in the potassium/argon ages of these minerals. Radiogenic argon may in fact provide an ideal tracer for volatiles during metamorphism, providing its solubility in various minerals and fluids can be specified as a function of pressure and temperature.

INTERPRETING THE PAST

One of the larger aims of geochemistry is to trace the way in which the observable features of the earth have developed since its formation. The rock record covers the last 3.5 to 3.7 billion years of the earth's 4.6-billion-year history. The record before 3.7 billion years ago has either been destroyed or not yet identified. The moon may be of some help in filling this gap, although this is not yet established. Several major questions remain unresolved, despite the presence of a considerable time record on earth; have the continents increased in area with time, or were they formed as part of an early terrestrial differentiation? Is the mantle now undergoing differentiation, or did this differentiation occur early in its history to be limited later to minor recycling? Was any form of plate-tectonic cycle operative more than a few hundred million years ago?

By analyzing strontium and lead isotopic abundances in mantle-derived igneous rocks from various ages, we can study the isotopic evolution of the mantle through time. The evidence so far suggests that at least the upper mantle has not acted as an infinite reservoir throughout time, but has undergone fairly clear-cut chemical changes.

The argument of early crustal development (big bang) versus continuous accretion has not been resolved, however, because recycling of some crustal material through the mantle (by seafloor spreading and subduction processes) appears to reconcile the isotopic data with a big-bang model. Thus, the solution of this problem may ultimately depend on the question of whether plate-tectonic processes operated continuously or sporadically throughout earth history.

Comprehensive studies of Precambrian terrains, combining geochronologic, petrologic, and isotopic approaches with field mapping, structural, and sedimentologic studies, will be required before the tectonic style of the Precambrian can be assessed. Much Precambrian terrain is composed of interrelated metamorphic and igneous rocks. Although the field geologist can map various identifiable units, interpretation of geologic history requires techniques for dating the units, determining the pressure–temperature conditions of formation, and establishing the original parentage of the metamorphic rocks. For example, the earliest Precambrian (Archean) shields are composed of a complex juxtaposition of metamorphosed volcanic belts and metamorphosed sedimentary-plutonic belts. The metavolcanic rocks are strikingly similar in composition to the near-trench volcanics of modern island arcs. Are there chemical asymmetries or polarities in the Archean volcanic rocks and pressure–temperature polarities in the Archean metamorphic rocks that would suggest subduction tectonics, and allow actual reconstruction of arc geometry? What is the time relationship of adjacent metavolcanic belts in the Archean? What is the general time scale for a Precambrian orogenic cycle? With the possibility of 1-million-year time resolution from the zircon uranium/lead technique, one can study Precambrian orogenies in as much time detail as Cenozoic orogenies. Did the orogenic belts develop synchronously on the different continents? Is there evidence in the terrains bordering the Archean volcanic belts for a "primitive" or original sialic crust to the earth? Or does a given orogenic cycle always start with mafic volcanics? How are the Precambrian ore deposits related, temporally and geographically, to the various igneous rocks? Can remnant oceanic crust (ophiolite complexes) be recognized? If so, the difference between the crystallization age and emplacement age may yield information about spreading rates and distances between accreting and converging plate boundaries.

The whole understanding of the early development of the earth and the initial stages of continental evolution is involved in these questions. Happily, it appears that straightforward techniques and approaches are at hand to provide many of the answers. The coming decade should see a great improvement in the understanding of early earth history.

SELECTED READINGS IN SOLID EARTH GEOCHEMISTRY

Abelson, Philip (ed.). Researches in geochemistry, Vol. 1. New York: John Wiley and Sons, 1959. 511 p.

Abelson, Philip (ed.). Researches in geochemistry, Vol. 2. New York: John Wiley and Sons, 1967.

Goldschmidt, V. M. Geochemistry. Alex Muir (ed.). London: Oxford University Press, 1958.

Mason, Brian. Principles of geochemistry. New York: John Wiley and Sons, 1966.

Rankama, Kalervo, and Th. G. Sahama. Geochemistry. Chicago: University of Chicago Press, 1956.

Wedepohl, K. H. (ed.). Handbook of geochemistry. New York: Springer-Verlag, 1969.

York, Derek, and Ronald M. Farquhar. The earth's age and geochemistry. New York: Pergamon Press, 1972.

Exogenic or Low-Temperature Aqueous Geochemistry

DEFINITIONS AND GOALS

Exogenic geochemistry includes processes that occur on the surface of the earth, within the atmosphere, biosphere, ocean, and terrestrial water bodies, and in the sediments and sedimentary rocks with which these bodies are in contact. Especially important are the chemical reactions and transfer of chemical species that occur at the boundaries (interfaces) between these distinct units; for example, between the ocean surface and the atmosphere, or between sediments and seawater or river water (Figure 4).

The processes that occur under such conditions almost invariably involve reactions of chemical species in systems composed of solids and water (heterogeneous aqueous systems) at relatively low temperatures and pressures. Exogenic geochemistry is thus differentiated from solid earth geochemistry, in which the reactions of interest occur at much higher temperatures and pressures.

The goals of studies in this area are to understand the chemical behavior of natural systems such as the ocean, atmosphere, rivers, and sedimentary rocks and to identify sources, sinks, and pathways, and to determine the evolution of chemical components migrating through these systems.

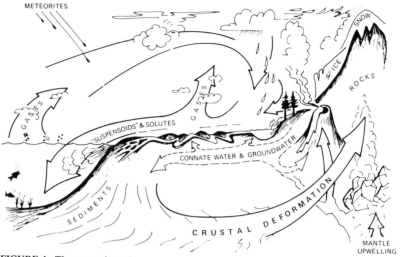

FIGURE 4 The exogenic cycle.

RECENT TRENDS

The most important trends in this field in the last few years have been the development of more realistic mathematical models for exogenic systems, and the application of the theory and methods of classical physical chemistry to natural systems.

The mathematical models, which are used to simulate the behavior of natural exogenic systems and to trace the migration of chemical species through them, provide a necessary theoretical framework that now allows geochemists to plan field investigations within a coordinated theoretical framework, to recognize areas where data are lacking, to consider interrelations between different parts of the exogenic cycle, thus providing an understanding of the whole cycle as well as simple descriptions of its isolated parts, and to predict the influence of man-made additions to the geochemical cycle at each point in the cycle.

The use of physical chemistry, principally thermodynamics, has made it possible for many natural reactions to be described quantitatively. One result of these studies has been the realization that complete chemical equilibrium is rarely attained in natural systems. Attention, therefore, has recently been focused on studies of the mechanisms and rates of reactions and on better definition of equilibrium states.

The increasing application of stable and radioactive isotopic techniques to geochemical problems in this area has made much of this progress possible. Ratios of stable isotopic species allow geochemists to follow the migration of chemical species through the exogenic cycle and to investigate the nature and rates of chemical processes that occur during each step of the cycle. Decay rates of radiogenic species are used for the same purposes.

FUTURE WORK

Accumulation of Needed Basic Data

Despite the increasing application of sophisticated chemical and mathematical methods to problems in exogenic geochemistry, an appreciation of the great complexity and diverse relations of the different parts of the exogenic cycle is just beginning to emerge. Large gaps in data, however, still exist.

It is clear, for example, that the chemical and mineralogical compositions of sedimentary rocks are major variables controlling the behavior of the exogenic cycle. Together with isotopic data, they also constitute observable boundary conditions that any potential mathematical model must satisfy. However, only one major study of the variation in chemical composition of a large volume of rocks through time is available: the work carried out by a Soviet group on the rocks of the Russian Platform. Similar large-scale, systematic, and interdisciplinary studies are needed for other continents, the ocean basins, and especially for Precambrian rocks.

There is also a noticeable scarcity of basic chemical data for the suspended load of streams and rivers. The major- and minor-element composition of the solid material carried by streams into the ocean is of basic importance to any considerations of the exogenic cycle, yet data are at present available for only two rivers, the Nile and the Mississippi. Little information exists on such other important problems as the amount and nature of organic matter and dissolved organic species in atmospheric precipitation, rivers, and groundwater; and the composition and amount of material transported into the oceans by glaciers and through the atmosphere.

Low-Temperature Chemical Reactions and Reaction Rates

Because it appears that equilibrium is rarely attained in the exogenic system, studies of low-temperature reaction rates and mechanisms are becoming increasingly important. Such studies require both field and laboratory investigations. Many natural reactions involve organisms such as bacteria. A greater emphasis will be needed to understand the effects of such biological activity on the transport and reaction rates of both inorganic and organic chemical species.

Knowledge of the rates of chemical reactions and of the biological and hydrological processes opposing or promoting equilibrium in natural systems will lead to the development of models for predicting transport rates and residence times within the atmosphere, ocean, lakes, and sediments for specific chemical species, including toxic substances of current concern, such as mercury, cadmium, and radioactive cesium and strontium.

Evolutionary Models

As data and modeling methods become adequate to explain the present exogenic cycle, it will become equally important to construct mathematical models that duplicate the system over periods of time in the past and for the future. Such models should allow us to understand the origin and evolution of the ocean and atmosphere and make possible predictions of the future state of the exogenic system. Such models should take into account both natural evolution and the effects of man-made contributions to the system.

Implications of the Plate-Tectonic Theory

The hypotheses of plate tectonics have several implications for exogenic geochemistry that must be evaluated. Two areas of particular importance are (1) the extent of recycling of original sedimentary rocks through the earth's crust and upper mantle and the effects of such recycling on the chemical processes of metamorphism and volcanism; and (2) the quantity of mantle-derived material, especially water and other volatiles, added to the ocean as a result of volcanic activity.

Chemical Changes in Sedimentary Deposits

Present understanding of chemical and physical changes in sediments after their deposition (diagenesis) is fragmentary. Of particular importance are the chemical changes that occur prior to metamorphism; for example, major alteration of clay minerals and organic compounds may occur during this time. Although laboratory time scales are short relative to geologic systems, causing difficulties in simulation, further experimental work is needed to elucidate the types of reactions involved in diagenesis.

Examples of Applications

• A better understanding of the behavior of exogenic systems is absolutely essential if man is to dispose of his waste products intelligently. The better the understanding of exogenic systems, their stability relations, and the pathways of individual chemical species through them, the better the effects of man-made additions to the systems can be predicted. Such predictions must underlie the establishment or modification of public policy regarding the addition of man-made materials to the environment.

• The study of exogenic geochemistry will provide data for understanding the properties and origins of an important class of ore deposits, the so-called stratiform metal deposits, which include present and potential economic sources of many industrially important metals, including iron, manganese, zinc, and copper.

THE OCEANS

Internal Reactions

More than 70 percent of the earth's surface is covered by oceans, which contain about 80 percent of the earth's free water (Table 2). This large reservoir is the site of many geochemically important reactions, both internally and at its interfaces with the atmosphere and the sediments.

A recent report issued by the Committee on Oceanography of the

TABLE 2 Water in the Exogenic Cycle

Reservoir	Water Mass $(10^{20}g)$	Percentage of Total
Ocean	13,700	79.6
Total pore water (including groundwater)	3,300	19.2
Ice	200	1.2
Lakes	2.3	0.013
Rivers	0.012	0.00007
Soil moisture	0.82	0.005
Atmosphere	0.13	0.0008

National Academy of Sciences, *Marine Chemistry* outlines the history, current status, and many of the outstanding problems encountered in studies of ocean chemistry. Here we shall touch lightly only on problems that are of particular importance to geochemists.

The 1960's might be described as the decade of equilibrium marine chemistry. Starting with Lars Gunnar Sillén's postulate that the ocean is, at least as a useful approximation, a solution in chemical equilibrium with a number of solid and gaseous phases, much progress has been made in determining the thermodynamic properties of the major components of seawater. As a result, the speciation and apparent equilibrium constants as functions of temperature, pressure, and salinity for the carbonate system, for example, are well known. Similarly, new experimental techniques for working in solutions of high ionic strength and species complexity, such as seawater, have allowed determination of ion pairing of the major dissolved species—sodium (Na^+), potassium (K^+), magnesium (Mg^{++}), calcium (Ca^{++}), sulfate ($SO_4^=$), bicarbonate (HCO_3^-) and carbonate ($CO_3^=$)—and development of a theoretical base for further work. Important work remains to be done, however, on the less abundant constituents, many of which are geochemically important. We are only just beginning, for example, to determine the forms in which transition metals occur in seawater.

Although the concept of the ocean as a chemical system in an equilibrium or near-equilibrium state has led to major improvements in the methodology of marine geochemistry and will undoubtedly lead to a still better understanding of speciation of dissolved components in the future, results obtained during the past few years suggest that deviations from complete heterogeneous equilibrium are the rule, rather than the exception.

The components most obviously out of equilibrium are those utilized by organisms. Both calcium carbonate and silica belong to this category and have been studied by geochemists for almost a century. In both cases, planktonic organisms fix these components at rates greatly in excess of the annual influx from rivers. Thus, if the ocean is not to be stripped, most of the biologically fixed material must redissolve in the water column or on the seafloor. The distribution of such components is, therefore, determined by the rates of fixation and dissolution, and by oceanic circulation. Many of the factors affecting the distribution of calcium carbonate and silica in the ocean and on the seafloor today have been identified, but there is little understanding of the mechanisms involved. It is one thing to know that the rate of dissolution of calcium carbonate increases with depth in the ocean, but quite another to estimate this rate solely from a set of physical and chemical parameters.

Preliminary experiments with the carbonate and silica systems show promise of leading to predictive models for the dissolution rates of calcite and opaline silica, respectively. Much more work will be necessary, however, before the present patterns are fully understood, let alone before their response to modifications of the boundary conditions is known. Also, studies of seawater to determine controls of saturation states, nucleation rates, and the degree to which seawater composition may be "pushed" by addition of chemicals, are greatly needed.

Information on most trace elements and organic compounds is at a far more primitive level. The ways in which their concentrations vary throughout the ocean, their biological activity, their speciation, and the processes that transfer these components from solution to solid phases in sediments are virtually unknown. This lack of knowledge is particularly unfortunate because transition and heavy metals, which belong to this group, are entering the ocean in large quantities as industrial wastes. As an illustration of this problem, man's global production of lead and mercury is approximately 5 and 20 times, respectively, the annual natural input of these elements to the ocean. International surveys, such as the Geochemical Ocean Sections Program for the International Decade of Ocean Exploration (GEOSECS) can provide information on the global distributions of a number of these elements. More extensive studies in areas of upwelling, where the fluxes are particularly great, are also needed. Further development of automated sampling and analyzing systems may yield many continuous vertical-concentration profiles to replace the few now available that are based on analyses of widely spaced samples.

As better analytical data have been obtained, increasingly more sophisticated advection–diffusion models have been developed to explain the observed vertical distribution patterns of seawater components. At present, it is not clear to what extent these models explain the real world and to what extent they are forced to fit the data by adjustment of unmeasured coefficients. Better estimates of coefficients such as those for eddy diffusion are needed, as well as more critical testing of the models with new data, particularly isotopic data.

All the models now being used, and implicitly the GEOSECS program, assume a steady-state ocean. Recent observations of highly variable deep-ocean currents cast some doubt on the validity of this assumption. Repeated occupation of oceanographic stations at intervals ranging from hours to years and use of the most precise analytical techniques available will test our confidence in time-independent models. In addition, geochemical studies of sediment cores should be expanded to evaluate better the steady-state assumption and the effects of glaciation on global environmental conditions over periods of hundreds to millions of years.

INTERACTIONS OF THE OCEAN WITH OTHER COMPONENTS OF THE EXOGENIC CYCLE

Atmosphere

Of the chemical processes occurring at the three interfaces separating the air–land–sea system, less is probably known about the air–sea interface. More attention has recently been paid to the chemistry of the air–sea interface as its role in environmental and geochemical problems has become more apparent. The solutions to many problems of air and water pollution, as well as a fuller assessment of the geochemical cycles of many substances depend on a better understanding of sea-surface phenomena.

Particle Exchange A better estimate of the global production rate of atmospheric sea-salt particles by bursting bubbles in the ocean is needed. Information on the size and frequency of bubbles in the sea under varying wind conditions is a prerequisite for making this estimate. In addition, studies are needed of the production of atmospheric particles by the mechanical tearing of waves, with particular emphasis on the size and number of particles produced.

There is a consensus that, on the basis of total sea-salt aerosol mass, chemical fractionation does not significantly affect the major-element composition of the marine aerosol. Evidence exists, however, that some elements show a tendency to be enriched on the smallest particles ($r < 1$ μm) produced by bursting bubbles. Further, investigations of the amounts and mechanisms involved in chemical fractionation, paying particular attention to the transfer of organic matter from bulk water to the sea surface and to the chemistry of film drops produced by bursting bubbles, are desirable. There is little information on the chemistry of the sea-surface microlayer and the effect of surface slicks, both natural and artificial, on particle and gaseous exchange. In general, rates and mechanisms of transfer of substances to the sea surface, both from above and below, are unknown. As an example of the importance of atmospheric deposits, it appears that a significant portion of the lead burden of surface ocean waters is derived from the atmosphere.

Gas Exchange The concentrations of atmospheric gases in the sea depend on pressure, salinity, and temperature conditions, and on reactions within the ocean. Most dissolved gases are not present in equilibrium concentrations, owing to kinetic effects at the sea surface. Current models generally assume turbulent water and air phases separated by a laminar layer, across which gas is exchanged by diffusion. The thickness of the laminar layer and the rate of gas exchange is dependent on sea state, wind speed, temperature, and sea-surface films, but not much is known about the relative importance of these factors. Recent work has suggested that transport of atmospheric gases in bubbles formed from breaking waves is also an effective means of transferring gases from the atmosphere to the ocean. The relative importance of bubbles versus classical molecular diffusion through the surface laminar layer should be evaluated.

The concentration of many of the minor gases in seawater above saturation values has been used extensively to describe mixing and advective processes in the ocean. Greater use of these gases as tracers will result from more accurate determinations of their natural solubilities. This is particularly true for xenon and krypton. An investigation of variations in the solubilities of the minor gases at different partial pressures should also be made.

Sediments and Interstitial Waters

Studies of interactions among seawater and minerals, both in suspension and at the seafloor, have proved to be one of the most exciting, and at the same time, frustrating aspects of marine geochemistry during recent years. An evaluation of these chemical interactions is necessary to understand the processes that modify the composition of seawater and sediment as originally deposited, the fluxes of chemical materials into and out of modern sediments, and the mineralogical sinks of elements, including pollutants, entering the ocean.

Solid Phases Laboratory experiments show that fine-grained silicates react rapidly with seawater. Isotopic measurements on pelagic sediments of terrigenous origin, and theoretical calculations, however, indicate that the mass transfer involved during reaction in the ocean is small, perhaps owing to armoring of the silicate grains by coatings.

Most studies of bottom sediments have dealt with the oxidized deposits that characterize pelagic areas. These sediments are rich in iron oxides, manganese oxides and hydroxides, as well as in a number of trace metals. In some instances the hydroxyoxide phases are dispersed in the sediment; in others they are concentrated in layers in the sediment; also they form concretions or nodules of potential economic value. In no case are the constituent phases well determined or the mechanisms by which the metals are deposited from solution understood. Laboratory studies utilizing radioactive tracers would seem a fruitful area for research in this field. Use of naturally fractionated radioactive daughter elements of the uranium and thorium decay series has proven invaluable in estimating rates of accumulation of pelagic sediments and ferromanganese nodules. Work on the fractionation processes should further increase the usefulness of these tracers.

Two complex aluminosilicate minerals, the zeolite phillipsite and an iron-rich montmorillonite, formed either by alteration of volcanic debris or by direct precipitation from solution, are common components in slowly accumulating deep-ocean clays. In addition, an authigenic (formed after deposition) iron-rich clay mineral has also been identified in shallow-water marine sediments. Sepiolite, a magnesium hydroxy-silicate, is abundant in some deep-sea sediments, but is rare in modern marine deposits. The influence of such minerals on the chemical composition of seawater is not completely known, nor is their effectiveness as sinks for trace elements.

"Reduced" sediments (usually defined operationally on the basis of their green to black color) cover a much smaller area of the seafloor than oxidized deposits, but are much more important volumetrically, and perhaps of greater interest geochemically. These deposits range from the shallowest estuarine muds to moderately deep-water silts and clays. Most of the suspended load of rivers ends up in deltas or in other bodies of reduced sediments.

Geochemically, reduced sediments are the least understood of any marine deposits. Their high contents of organic matter, complex lateral variations, and associations with high biologic productivity and the zone of mixing of rivers and the ocean constitute formidable obstacles to the recognition of systematic geochemical characteristics or patterns of behavior. Yet these sediments appear to be the major sink for silica, sulfur, and heavy metals introduced to the ocean in solution, and may well dominate the geochemical behavior of many trace elements. Because such deposits form the long-term sinks for most of man's wastes, development of a systematic body of geochemical theory is particularly urgent. Careful determinations of accumulation rates, compositions and exchange properties of both detrital phases and those formed shortly after deposition would do much to explain the effects of such deposits on the chemistry of seawater.

Aqueous Phase Studies of interstitial waters from both nearshore and deep-sea marine sediments have revealed vertical concentration gradients in a number of chemical species. In general, Mg^{++} is depleted in interstitial waters with respect to normal seawaters; Ca^{++} and K^+ are either enriched or depleted, and Na^+ appears constant or depleted. As far as the anions are concerned, HCO_3^- is generally enriched whereas $SO_4^=$ is depleted. Silica (SiO_2) is usually enriched, but concentrations are generally lower than the saturation value for amorphous silica. Reduced and oxidized sediments appear to have different pore-water characteristics. Chemical gradients in pore waters of reduced sediments are more pronounced and exist for more species than in oxidized sediments. The greater the organic matter of a sediment layer, the larger the gradients for $SO_4^=$ and Mg^{++}. The influence of mineralogy on pore-water composition is still uncertain.

A number of attempts have been made to explain these observations; for example, it has been suggested that Mg^{++} from pore waters replaces Fe^{+++} in clay minerals, with the simultaneous reduction of Fe^{+++} and precipitation of Fe^{++} as iron sulfide accounting for the covariance of $SO_4^=$ and Mg^{++}, but more work is obviously necessary.

Because most of the data collected from the deep sea was obtained by extraction of pore waters at shipboard temperatures and low pressures and not *in situ*, the dissolved SiO_2 and K^+ data are suspect. This fact may explain some of the K^+ gradients noted. Also, efforts to obtain accurate analytical data for the complete chemistry of interstitial water profiles are just beginning. Data for elements in interstitial waters should include concentrations and profiles of minor, trace, and nutrient elements, organic substances, gases, and isotopes, as well as major elements measured on samples collected *in situ*. Interstitial water studies are needed on sediments from all water depths. The deeper sediments include most of the seafloor, and their interstitial water compositions must be known if global estimates of element fluxes are to be made. The nearshore zone must also be studied because it is a region of major environmental concern as well as a complicated zone of transfer of elements into the sediments and out to the greater mass of the ocean.

A major impediment to the synthesis of models for describing interstitial water data is the dearth of information on the chemistry and detailed mineralogy of the sediments from which the interstitial waters have been extracted. Detailed mineralogical analyses will entail a complex and expensive array of analytical equipment.

Evaporites (chiefly salt and gypsum deposits) The present is not a period of prominent evaporite deposition. The small evaporite basins that do exist, however, provide important clues to processes responsible for the great evaporite deposits of the geologic record. The recognition of contemporaneous dolomite [$CaMg(CO_3)_2$] in several modern evaporitic environments, for example, is one of the important discoveries of the past decade. Before this discovery, many geochemists contended that all dolomite was formed by postdepositional processes. Further work on the actual reactions involved, making use of stable isotopic and trace element abundances, should further clarify the genesis of the ancient deposits.

Igneous Material About one third of the seafloor consists of basalt, or of basalt thinly covered by sediments. This material is continually augmented at a rate of about 2 km^2 per year by new basalt emplaced at midocean ridges. Recent examination of sediments along midocean ridges, of seismic-velocity changes in the top of the oceanic crust with increasing distance from ridge axes, of the effect of chemical alteration on the seismic velocity of basalt, and of chemical changes measurable in basalt as a function of its age, all suggest that basement in the ocean basins may be an important component of the exogenic geochemical system; the fluxes of a number of elements from basalt to the ocean and vice versa may be as great as the freshwater input. Knowledge of basalt–seawater interaction is, therefore, important to balance the exogenic geochemical cycle. Because the basalt derived from the mantle has distinctive trace element and isotopic characteristics, it is possible to establish in some cases the origin of trace components in marine sediments. Iron-rich deposits from the East Pacific Rise in the southeast Pacific, an area of very slow terrigenous sedimentation, have lead isotopic ratios characteristic of basalt but strontium isotopic ratios and rare-earth patterns more reminiscent of seawater. Evaluation of the origin of such iron-rich sediments is an important goal for marine geochemists in the immediate future, particularly because many metal-rich sedimentary rocks on land have compositions and geologic settings similar to the metalliferous deposits.

The quantitative importance of interactions between basalt and seawater in the exogenic geochemical cycle depends on the depth within the basalt at which the interactions take place. Deep-basement samples to be collected by the drilling ship *Glomar Challenger* as part of the Deep Sea Drilling Project of the National Science Foundation will indicate some measure of this depth of alteration.

FRESH WATERS AND NONMARINE SEDIMENTS

The geochemistry of terrestrial waters is primarily concerned with reactions that occur in relatively dilute aqueous solutions. The chemical composition of these waters varies considerably with geographical location and time. The composition is dependent on factors such as the minerals with which the water has been in contact, the duration of contact, the amount of water involved and other variables such as pH, temperature, and biological activity. Reactions between water and the various minerals in igneous rocks are important, but sedimentary rocks and soil are apparently the primary sources of most of the dissolved material in surface waters.

Perhaps the most significant advance in this area in the past 10–15 years has been the application of thermodynamic principles to the understanding of the chemical composition of fresh water. The use of various types of equilibrium calculations has led to a much better understanding of the reactions between fresh water and rock minerals. This technique has been particularly successful in the study of limestone hydrology and silicate-mineral dissolution; considerable additional work in this area, however, is still needed. Many of the more complicated reactions involving mineral dissolution proceed too slowly for complete equilibrium to be attained. This is especially true for river systems. Thus, the development of a broad theoretical framework (a series of transfer functions) for integrating the effects of bacteria, organic matter, and armoring of materials, on reaction rates is a necessary next step. Studies of reaction kinetics will necessarily involve more detailed investigations of reaction mechanisms.

To a large extent the increase in research in lake and river chemistry has been a result of pollution in these bodies of water. As in the case of other geochemical systems that can be affected by man, there has been a realization that natural geochemical cycles must be understood before what happens to man's effluents in lake and river systems can be effectively understood and controlled.

Except for the major cations, few data are available for the cation exchange capacity or adsorptive properties of suspended river sediments. Particle–solute interaction in rivers may be rapid compared to solid–liquid interactions in groundwater, lakes, and the ocean. Kinetic studies are thus particularly important in river systems.

Rivers and glacial ice together carry more than 90 (and probably more than 95) percent of the dissolved and particulate material transferred from the continents to the ocean. The concentrations and rates of transfer of major anions and cations by rivers have been estimated on the basis of few measurements, often with little consideration being given to flow conditions at the time of sampling.

The size of the contribution of ice-transported continental debris to the marine geochemical cycle is highly controversial. Estimates range from less than 1 to more than 10 percent of the suspended load of rivers. If the high figure is correct, all preceding calculations of terrestrial input to the ocean must be revised. Both the quantity and characteristics of glacial debris supplied to the ocean should be determined.

A systematic survey of the trace element contents, as well as the concentration and composition of suspended matter, in the 20 major rivers that discharge 90 percent of the runoff to the ocean is of fundamental importance. Even now, such a survey will require correction because of man's activities, particularly in agriculture, if past rates of erosion are to be estimated. Concurrently with such a survey, determinations should be made of the speciation, and, for elements like sulfur and carbon, the isotopic composition, of dissolved trace elements in rivers, in estuarine mixing systems, and in near-shore waters of the ocean. Initial investigations of artificially produced radioactive elements and of heavy metals in industrial wastes point to important changes in complexing of transition and heavy metals and in interactions between solid and dissolved species in the zone of mixing of river and ocean waters. There is evidence that certain trace elements are important in controlling biological systems in fresh water. For example, cobaltous ion is present in vitamin B_{12}, and this substance may control species composition of planktonic algae. Molybdenum may affect the general productivity of surface waters, inasmuch as it is involved in nitrogen metabolism. Knowledge of organically bound constituents in these waters is minimal, and the extent of organic-metal complexing is largely unknown.

Organisms in fresh waters, particularly microorganisms, may significantly affect the concentration of many substances, as they alter reaction rates and equilibrium concentrations. Complex organic materials such as "humic" and "fulvic" acids and other polymers, which rapidly form in the water column from biologically derived substances, warrant study. Considerable additional information is required before the extent of the influence of biological activity on chemical equilibria and geochemical cycles in surface water can be ascertained.

More mathematical modeling of freshwater systems is badly needed. Application of such models has already been made, for example, to the transport and deposition of radioactive isotopes in lake waters and to metal-organic complexes in freshwater systems.

In addition to marine studies, pore-water analyses from freshwater sediments and mathematical modeling of the resulting distribution patterns are critical to our understanding of element cycles in nature. Of particular interest are the cycles of biologically active elements, such as carbon, nitrogen, phosphorus, and the importance of pore-water–sediment systems as sources or sinks for these elements.

Knowledge of the fate of metal and organic pollutants in surface waters is patchy. The mechanisms for the retention of the various pollutants in soils and sediments, the capacity of soils and sediments for these pollutants, the rates at which these processes operate—all are virtually unknown. Again, an understanding of reaction kinetics and complexing is critical. In studying and understanding the geochemistry of surface-water systems the need for interdisciplinary research among chemists, geologists, meteorologists, hydrologists, biologists, engineers, and water quality personnel becomes obvious.

SUBSURFACE WATERS

Approximately 20 percent of the water in the hydrosphere is contained in the pores of sediments and sedimentary rocks—a reservoir second only to the ocean (Table 2). These waters are variable in concentration, ranging from fresh potable waters to dense brines. The waters initially enter rock reservoirs via streams, percolate downward from soil horizons, or are trapped in the sediment pores during deposition. A very small percentage of them may be primary water from the mantle of the earth.

The deeper waters may circulate and mix beneath continents and enter the ocean beneath the drowned margins of continents. Surprisingly little is known about the composition of these continentally derived waters and their flow into the ocean. Perhaps 10 percent of the water annually entering the ocean follows this route. As exploration for oil and gas on continental shelves increases, some effort should be made to obtain uncontaminated samples of these waters during drilling, to assess their flow rates, to determine their chemical compositions, and to evaluate their effects on mineral changes in the sediments.

Drilling along some continental shelves (such as the Gulf Coast of the United States) has also revealed contrasting compositions of fluids in shales and sandstones. These waters probably result from seawater-mineral interactions during burial and from differential transport of dissolved constituents between the shales and sands. The details of the chemical processes that fractionate the original waters are little understood, as are the transport processes and fluxes involved. Studies of these waters could shed light on problems of diagenesis, fluxes of materials from subsurface to surface, and genesis and migration of hydrocarbons.

Subsurface waters within the continent are the source of several percent of the water used in the United States. Most of this water is derived from rain that has percolated downward through porous and permeable rocks and is susceptible to contamination by pollutants. Although much has been learned about the sources of constituents in these waters from equilibrium and material-balance studies, mathematical models need to be derived for these flowing systems in which reaction rates of minerals, bulk and diffusional transport of dissolved species, and equilibrium states are evaluated. These models are urgently needed because man continues to add much NO_3^- and $PO_4^{\equiv}$ to subsurface waters. Bacteria strongly influence reaction rates in percolating waters; consequently organic and inorganic geochemists and biogeochemists will find this an important field for integrative studies. Some work of this nature is already in progress, such as studies of biochemical reactions occurring in waters flowing through waste ore piles and sanitary landfills; however, more work is needed.

The distribution and genesis of chemical species in thermal waters and subsurface brines is of continuing interest. Thermodynamic modeling has been of help, but kinetic and transport models are lacking. Use of the wide spectrum of ion-sensitive electrodes now available may prove to be of help in determining *in situ* species distributions in brines. Studies of deep thermal waters will undoubtedly provide further data on transport of metals in solution and on their deposition. Computer modeling of these systems has been particularly helpful and should continue to be a powerful approach.

SOIL AND SOIL WATERS

Little is known about the details of the chemical processes operating in the soil zone. Many of the processes involved in soil formation, however, are the same as those that control the composition of surface water (although this has not been realized extensively by workers in these areas). Most terrestrial precipitation falls on soil surfaces and most of the runoff that enters streams and lakes has had contact with the soil. A large fraction of stream and lake water has often spent some time as soil moisture. Reactions of precipitation with soil are often quite rapid, so that direct runoff has a considerably higher dissolved-solid load than the original precipitation. Carbon dioxide derived from bacterial decay of organic matter is a particularly aggressive weathering agent. Many chemical changes take place when living organic matter is converted to soil humus; these changes are important in the carbon/nitrogen/phosphorus cycles. Analyses of gases in soil waters are scarce. Material balances are lacking because, for example, the mass of cyclic salt deposited by rain and dry fallout is not well known. Processes by which soil water moves down to the water table are also poorly understood. The soil zone and reactions occurring there are of immediate interest because this zone is a way station for many shallow-groundwater pollutants.

Problems of soil pollution have received relatively little attention. Apart from studies of fertilizer retention and availability, the mechanisms by which soils retain various elements, the reaction rates involved, and the capacity of soils to hold these elements are incompletely known. However, there is a growing demand for geochemical surveys of large areas in order to define existing levels and patterns of trace substances in the soils, so that subsequent pollution can be recognized. In this context, comprehensive studies of entire watersheds in a number of climates should be undertaken.

In general, there has been too little interaction between soil chemists and geochemists. Increased contact between these groups should prove beneficial to both.

THE ATMOSPHERE

Research in atmospheric chemistry on local, regional, and global scales has increased tremendously in the past 10 years. To a large extent, this has resulted from concern that man can inadvertently modify the atmosphere on all scales as a result of his technological achievements. This in turn has led to the realization that a complete understanding of the reactions and cycles of natural and man-made substances in the stratosphere and troposphere is necessary to assess man's impact. Increased interest has also stemmed from a realization that the atmosphere is important in many geochemical cycles, and an awareness by atmospheric scientists of the importance of atmospheric chemistry in studying cloud and rain formation, atmospheric mixing processes, and changes in the global radiation budget.

Trace Gases

The distribution and cycle of carbon dioxide in the atmosphere has been intensively studied. The long-range climatic effects of the accumulation of carbon dioxide in the atmosphere because of the burning of fossil fuels is still, however, subject to controversy. Continued monitoring of atmospheric carbon dioxide is mandatory.

Until the late 1960's, man's activities were generally believed to be the primary source of atmospheric carbon monoxide. It is now believed that biological activity in the ocean and atmospheric reactions involving oxidation of atmospheric methane and other organics by the hydroxyl (OH) radical (and perhaps other species) far outweigh man-made carbon monoxide. Details of these oxidation mechanisms are still sparse, and accurate kinetic data for the various steps in the reaction chains are especially needed. Very few data are available on the vertical distribution of carbon monoxide, particularly in the stratosphere. Soil has the capacity to remove all atmospheric carbon monoxide, but recent vertical profiles suggest that reactions in the lower stratosphere, for example, $CO + OH \rightarrow CO_2 + H$, are efficient sinks as well. The nature of these reactions and their relative importance in the carbon monoxide system need further study.

The general features of the atmospheric sulfur cycle are relatively well understood, although there are notable gaps. Although it now appears certain that ocean water is not a major source of hydrogen sulfide (H_2S), the recent suggestion that dimethyl sulfide [$(CH_3)_2S$] is produced in the sea should be investigated. The total quantities of naturally occurring H_2S produced over the continents and coastal areas are unknown. Virtually nothing is known about the ocean as a source or sink for atmospheric sulfur dioxide. More detailed measurements of tropospheric and stratospheric H_2S and sulfur dioxide concentrations are needed to determine sources and atmospheric residence times of these species. In distinguishing between various sources of sulfur, both natural and pollutant. $^{32}S/^{34}S$ ratios have been of some help, but better means are needed. Continued studies of the conversion rates of gaseous water and sulfur dioxide to particulate sulfates by photochemical and catalytic oxidation are needed. Investigation of volcanic production of sulfur compounds, especially related to particle concentration in the stratosphere should be continued.

The global natural production rates of atmospheric nitrogen oxides (NO_x) and ammonia (NH_3) are poorly known, as are their natural sinks and removal mechanisms. Nitrous oxide may be a major source of NO_x in the upper troposphere and stratosphere via reaction with atomic oxygen (O). Reliable vertical and horizontal profiles of nitrous oxide (N_2O), nitric oxide (NO), nitrogen dioxide (NO_2), and NH_3 are needed, especially in the stratosphere. Because of their influence on particle formation in the atmosphere, the mechanisms and reaction rates for the conversion of NO_x and NH_3 to nitrate and nitrite need to be determined.

Stratospheric ozone (O_3) is produced primarily at 20–30 km altitude by the photochemical dissociation of molecular oxygen (O_2) and subsequent reaction of the atomic oxygen with O_2 and a third body. Ozone absorbs ultraviolet radiation below 3100 Å and thus provides an important protective blanket against radiation harmful to living cells. Ozone participates in a number of stratospheric reactions with nitrogen oxides and such species as OH and the hydroperoxyl radical (HO_2), whose source is stratospheric water vapor. Because of concern for the possible effects of supersonic transport (SST) emission of H_2O and NO_x on the ozone budget in the upper atmosphere, there has been a dramatic increase in research on chemical processes in the stratosphere and mesosphere during the past few years. However, much more information about the concentration, production, and removal rates of such reactive species as NO, NO_2, H_2O, OH, HO_2, H, and O in the upper atmosphere is needed before a complete evaluation of SST effects can be made.

Considerable interest has recently been shown in the atmospheric chemistry of methane and other low molecular weight hydrocarbons (e.g., ethylene), partly because of the realization that these substances may be a major source of atmospheric CO, as mentioned previously. Methane itself is apparently produced chiefly through anaerobic decomposition of organic matter in swamps, lakes, marshes, and paddy soils. There is also evidence that the surface of open ocean waters is generally supersaturated with methane (relative to atmospheric methane) although the ocean's importance as a global methane source has not been evaluated. Knowledge of the vertical distribution of methane is fragmentary and contradictory, although there is a definite decrease in concentration in the stratosphere. More information is needed on the chemical reactions and rates leading to the destruction of atmospheric methane. Our knowledge of the atmospheric distribution of other naturally produced hydrocarbons is extremely limited. We need to determine the role of the heavier hydrocarbons as possible sources for atmospheric methane and carbon monoxide and their importance in the formation of atmospheric particles.

It is generally believed that a large fraction of the chlorinated hydrocarbons found in the marine environment are transported via the atmosphere, but there is little information available on the atmospheric concentration or distribution between the gas and particulate phases of chlorinated hydrocarbons.

Aerosols

Troposphere Evidence now exists that tropospheric aerosols fall into three general classes. Although the number, size distribution, and chemical composition of aerosols near the earth's surface vary considerably with time and geographical location, the total aerosol abundance is relatively constant above about a 5-km altitude. Atmospheric particles in this region are generally referred to as the background aerosol. Their importance lies in their possible influence on the global radiation budget and on cloud formation. However, chiefly because of the unavailability of proper sampling and analytical techniques, almost nothing is known about the particle-size distribution, chemical composition, production mechanisms, or detailed spatial and temporal variability of the background aerosol.

Particles below the 5-km altitude over continents are referred to as the continental aerosol and are probably a major source for the background aerosol. The primary sources for the continental aerosol are believed to be crustal weathering, gas-to-particle conversion reactions, and man-produced emissions. Production rates from any of these sources are not well known. More information on the variation of the size distribution and chemical composition of the continental aerosol is also needed.

Particles found below about 5 km over the ocean are generally referred to as the marine aerosol. They are largely composed of sea-salt particles superimposed on the background aerosol. The abundance, size distribution, and major-element chemical composition for sea-salt particles $\geqslant 1$ μm radius are fairly well known. Few data are available for smaller particles, either with respect to production mechanisms or the parameters above.

Stratosphere Information on particles in the stratosphere is meager. Stratospheric aerosols have residence times of 1–2 years, considerably longer than tropospheric aerosols. The most prevalent anion in stratospheric aerosols is $SO_4^=$, but the number, size distribution, and detailed chemical composition of stratospheric aerosols, as well as the spatial and temporal variations of these parameters are generally not known. In particular, the composition and size distribution of particles < 0.1 μm radius are completely unknown. Details of the mechanisms and rates of gas reactions producing stratospheric particles are largely lacking. The high concentration of sulfate-containing particles with $r = 0.1–1.0$ μm in the 18–20-km altitude region is now well established. Although the source of the particles is still subject to some controversy, it appears that they may result from a pulse-type injection mechanism, most likely volcanism. More work should be done to test this hypothesis. New analytical and sampling techniques will probably be necessary before a complete understanding of stratospheric aerosols is obtained.

Interaction of cosmic rays with the upper atmosphere produces a number of radioactive nuclei, both directly and by means of secondary reactions, which have been useful as tracers in elucidating geochemical processes over various time spans. For example, relatively long-lived cosmic-ray-produced nuclides, such as ^{14}C, ^{26}Al, and ^{10}Be, have been used to determine the chronologies of oceanic and fresh-water sediments. Nuclides with half-lives of a few hundred years, such as ^{39}Ar and ^{32}Si (as well as longer-lived ^{14}C), have been used in rudimentary studies of oceanic circulation, whereas short-lived nuclides such as ^{3}H, ^{7}Be, ^{22}Na, ^{35}S, and ^{38}Cl, have been used in studies of atmospheric washout and short-term mixing processes in the ocean. The use of these radioactive tracers will continue to be of great importance to geochemistry.

Removal Processes

Atmospheric removal rates for most chemical species in the gas and particulate phases are not available. Knowledge of atmospheric mass transport and deposition on both local and global scales is fundamental to understanding the migration of these species through the exogenic system. Present information suggests that approximately 75 percent of the particulate matter in the atmosphere is removed by precipitation and the remainder by dry fallout, but this estimate is based on few data. These figures probably vary widely with such factors as geographical location, season, chemical composition, and size of the particles. Recent investigations using the ratio of the atmospheric radon daughters $^{210}Bi/^{210}Pb$—corrected for a secular equilibrium component of ^{210}Pb, ^{210}Bi, and ^{210}Po from soil surfaces—show an average residence time of particles in the lower troposphere of 4–7 days. This is shorter than previous estimates.

In the past few years research interests in precipitation chemistry have shifted from an emphasis largely on measuring concentrations of substances in precipitation, that is, how much material is transported to the earth's surface, to the problem of how material is scavenged from the atmosphere by precipitation. Below-cloud washout by rain of particles $> \sim 1$ μm radius can now be predicted and explained reasonably well. Theoretical treatments for removal of submicron particles and both reactive and nonreactive gases are inadequate, although progress is being made.

In-cloud rather than below-cloud scavenging of particles and gases is probably more important on the global scale; however, our understanding of the various processes at work in in-cloud scavenging, such as nucleation, diffusiophoresis, thermophoresis, Brownian motion, electrical effects, and impaction is fragmentary. Some of these processes can be predicted individually, but it has been impossible to construct an all-inclusive theoretical model to explain in-cloud scavenging. Knowledge of the effects of natural and man-made substances on such meteorological processes as cloud and rain formation is minimal.

ADVANCED DIAGENESIS

The changes that a sediment undergoes between deposition and eventual uplift and erosion constitute a broad realm of ignorance. We have only begun to understand some of the chemical and mineralogical processes operating during this phase of the exogenic sedimentary cycle.

Carbonate sediments begin as a metastable mixture of aragonite, calcite, and a spectrum of magnesium calcites; ancient carbonate rocks are composed principally of calcite and dolomite. Although the nature and genesis of modern and ancient carbonate sediments have been studied intensively, little is known about the chemical and mineralogical processes that lead to stabilization of these sediments. Research efforts in Barbados, Bermuda, Jamaica, and Florida have shown that the Pleistocene is the key to the past, if one wishes to understand carbonate mineral changes when they are exposed to freshwater above and below the water table, as well as in the brackish zone of mixed fresh- and saltwater. Studies of carbonate sediments collected during the Deep Sea Drilling Project suggest that changes of carbonates in seawater to stable solid species are slow, but do occur to a significant extent. Investigations of the chemistry (including isotopic composition) and mineralogy of Pleistocene and slightly older carbonate sediments are just beginning to shed light on the processes that lead to the stabilization of carbonate sediments. This area of research is likely to be rewarding in the future.

Some of the outstanding problems include the influence of the brackish water zone on diagenesis, particularly dolomitization; the rates of mineralogical transformation; the changes in sediment fabric and chemistry that occur in the deep groundwater environment; the fluxes of elements, particularly magnesium, into and out of carbonate sediments during diagenesis; and the changes of chemistry, mineralogy, and fabric that accompany silicification of carbonates.

Silicate sediments experience changes similar to those affecting carbonates. A number of studies have been made of early postdepositional reactions of terrigenous clay minerals with seawater; in general, except for ion exchange, chemical and mineralogical changes are small. Recent investigations of nearly continuous sequences of shaly sediments on the United States Gulf Coast, have shown that mixed-layer montmorillonite-illite clays are converted with burial diagenesis to more illitic clays; they gain potassium and aluminum, while losing silicon. This transformation has been documented by chemical analysis, X-ray diffractometry, and potassium/argon age-dating studies. Changes in the chemistry and mineralogy of shales as a function of geologic time have also been noted. These changes, which depend in part on depth of burial, are apparently only half completed in 30 million years in the average shaly sediment. Silicate sediments, like carbonates, appear to start out as a metastable mixture of phases and are transformed during burial to a chemically stable assemblage.

Investigation of the chemistry of solid and fluid phases in "piles" of silicate rocks is a major area of research in advanced diagenesis. Questions like the following may be answered using this approach: Does the mineral chlorite, a magnesium-aluminum silicate, form during this process? What is the spectrum of reactions involving mixed-layer clays? What are the compositions and fluxes of fluids and dissolved species involved? What are the magnitudes of the fluid and solute fluxes reaching the earth's surface during this phase of the sedimentary cycle?

Detailed investigations, using modern analytical techniques, of the mineralogy and chemistry of shales, of trace minerals in limestones, and of cements in sandstones are also needed. In most such investigations, more effort should be directed toward isolating and identifying the solid phases. Some experimentation on silicate systems under varying environmental conditions may help in isolating the important variables. An ultimate aim of the geochemist is to understand post-depositional changes in sediments, so that he can reconstruct their initial characteristics.

SUBDUCTIVE MASS TRANSFER AND CYCLING

Before the development of the plate-tectonics hypothesis the cycling of elements in the earth-surface system was assumed to be a virtually closed system. It now appears that mantle derivatives are being introduced at spreading midocean ridges, and that the seafloor with its layer of sediment, is re-entering the interior of the earth by subduction. The question arises as to the openness of the exogenic cycle, and the degree of importance of interactions between subcrustal and surficial layers of the earth.

If the plate-tectonics hypothesis is adopted as a working model for the moment, subduction zones lie vertically below volcanoes. The sediments and weathered basalt at the top of the downgoing lithospheric slab melt first because of their high water content. Many andesitic island arcs are probably formed from this material, allowing it little opportunity to react with the upper mantle. In addition, reflection seismic profiles suggest that some of the sediment on the subducted plate has been scraped off and accreted to the ocean side of the island arcs; rates of continental accretion over geologic time are equivalent to removal of about 1 or 2 km from the top of subducting lithosphere.

The loci of earthquake epicenters along Benioff (subduction) zones show that the subducting slab is coherent and rigid enough to fracture at depths of at least 700 km and may sink even further into the mantle.

Apparently little contamination of the upper mantle results from recycling of exogenic material. Recent measurements of noble-gas abundances in quenched margins of basalt erupted along midocean ridges support this conclusion. The noble-gas abundance distributions are primordial, similar to solar or meteorite patterns, but are quite unlike atmospheric or seawater patterns. Noble-gas patterns more reminiscent of the atmospheric pattern are found in seamount volcanic rocks, suggesting that some exogenic material may be entering the deeper mantle.

It appears that subduction zones are of geochemical interest largely because of the reactions and phase changes that occur in them, rather than because they mark a region of material transfer from one geochemical system to another. The possibility that ore bodies in andesitic terrains, among others, result from the distillation of subducted deep sea metalliferous sediments has obvious economic implications.

EVOLUTION OF THE EXOGENIC SYSTEM

The chemistry, mineralogy, relative proportions, and total masses of sedimentary rocks provide a basis for interpreting the history of the system for the past 3.5 billion years. Unfortunately, few of the basic data of the chemistry and mineralogy of sedimentary rocks have been collected. Only for the Russian Platform, and then mostly for Phanerozoic rocks (0–600 million years old), have these parameters been measured systematically.

Stable isotopes provide a powerful tool for evaluating the chemical stability of the exogenic system. The δ^{34}S content of evaporites, the δ^{13}C and δ^{18}O content of carbonates, the δ^{18}O content of cherts, the δ^{13}C content of organic sediments, the ^{87}Sr/^{86}Sr ratio of marine fossils, and the δ^{34}S content of sedimentary pyrite (FeS_2), are known to have changed through geologic time. Much of this information, however, is restricted to Phanerozoic rocks. More data are needed for Precambrian rocks. Stable isotopic data for sediments whose ages are known to within 5–10 million years are needed to evaluate the degree of correlation between different isotopes and to define changes and rates of change through geologic time. The data now available are integrated over relatively large intervals of time.

Significant mass transfer of elements into and out of the ocean by low-temperature and intermediate-temperature reactions between seawater and submarine basalts is implied by geochemical and seismic studies of Layer II, the seismic unit beneath the seafloor sediments. Such reactions, when superimposed on the apparent variability of seafloor spreading rates through geologic time, could significantly influence the chemistry of the ocean.

Subduction zones may be sites of transfer of materials into the mantle but they are also sites of chemical and mineralogical reactions that affect the exogenic cycle. Studies of such reactions, when combined with geophysical observations, may lead to more quantitative estimates of the degree of mobilization of elements in subduction zones.

The cycles of carbon, nitrogen, and phosphorus deserve special attention. Phosphorus in sedimentary rocks occurs in organic matter and in mineral form, but its partitioning between these phases as a function of geologic time is unknown. Little is known of carbon and nitrogen contents of Precambrian rocks. Collection of appropriate data will help us to recognize variations in rates of photosynthesis and organic productivity through geologic time and, perhaps, draw attention to the factors that control these variables.

Once the necessary data are obtained, mathematical models of the exogenic sedimentary system can be derived to estimate fluxes between different reservoirs and to determine the mechanisms by which mass transfer takes place. Steady-state models for some elements have already been constructed. Although these models suffer from insufficient data, they do provide a baseline against which excursions can be evaluated.

Transient models of the system are also needed. Such models can be tested against the geologic record to provide reasonable constraints on variations in the chemistry of the earth's surface environment over geologic time. The importance of these models in assessing the effects of pollutants on the environment is obvious. More sophisticated models should evaluate other reservoirs, such as the earth's mantle and extraterrestrial space, and the fluxes associated with them.

Data on the surface environment of other planets are needed for their own sake and to understand the early history of the earth, particularly of the atmosphere and hydrosphere. These histories will be invaluable for interpreting conditions before the age of the oldest sedimentary rocks so far discovered on the earth—3.5 billion years.

SELECTED READINGS ON EXOGENIC OR LOW-TEMPERATURE AQUEOUS GEOCHEMISTRY

Berner, R. A. Principles of chemical sedimentology. New York: McGraw-Hill, 1971. 240 p.

Broecker, W. S., and V. M. Oversby. Chemical equilibria in the earth. New York: McGraw-Hill, 1971. 318 p.

Butcher, S. S., and R. J. Charlson. An introduction to air chemistry. New York: Academic Press, 1972. 241 p.

Cadle, R. D. Particles in the atmosphere and space. New York: Reinhold Publishing Co., 1966. 226 p.

Committee on Oceanography. Marine chemistry: A report of the Marine Chemistry Panel. Washington: National Academy of Sciences, 1971. 61 p.

Detwyler, T. R. Man's impact on the environment. New York: McGraw-Hill, 1971. 731 p.

Garrels, R. M., and C. L. Christ. Solutions, minerals, and equilibrium. New York: Harper and Row, 1965. 495 p.

Garrels, R. M., and F. T. Mackenzie. Evolution of sedimentary rocks. New York: W. W. Norton and Co., 1971. 397 p.

Hood, D. W. (ed.). Impingement of man on the oceans. New York: Wiley-Interscience, 1971. 738 p.

Horne, R. A. Marine chemistry: the structure of water and the chemistry of the hydrosphere. New York: Wiley-Interscience, 1969. 568 p.

Joint Oceanographic Institutions for Deep Earth Sampling. Initial reports of the Deep Sea Drilling Project. University of California, Scripps Institution of Oceanography. Prepared for the National Science Foundation. Washington: Government Printing Office, 1969–1973.

Junge, C. E. Air chemistry and radioactivity. New York: Academic Press, 1963. 382 p.

Massachusetts Institute of Technology. Inadvertent climate modification: Report of the study of man's impact on climate. Cambridge: M.I.T. Press, 1971. 308 p.

Massachusetts Institute of Technology. Man's impact on the global environment: Report of a study of critical environmental problems. Cambridge: M.I.T. Press, 1970. 319 p.

Matthews, W. H., W. W. Kellogg, and G. D. Robinson. Man's impact on the climate. Cambridge: M.I.T. Press, 1971. 594 p.

Matthews, W. H., F. E. Smith, and E. D. Goldberg (eds.). Man's impact on terrestrial and oceanic ecosystems. Cambridge: M.I.T. Press, 1971. 540 p.

Ocean Science Committee. Marine environmental quality: Suggested research programs for understanding man's effect on the oceans. Washington: National Academy of Sciences, 1971. 107 p.

Riley, J. P., and R. Chester. Introduction to marine chemistry. London: Academic Press, 1971. 466 p.

Riley, J. P., and G. Skirrow (eds.). Chemical oceanography (2 vols.). New York: Academic Press, 1965. 712 p. (vol. 1); 508 p. (vol. 2).

Strahler, A. N., and A. H. Strahler. Environmental geoscience. Hamilton, California: Hamilton Publishing Co., 1973. 511 p.

Stumm, W. (ed.). Equilibrium concepts in natural water systems. Washington: American Chemical Society Advances in Chemistry Series, 1967. 344 p.

Stumm, W., and J. J. Morgan. Aquatic chemistry: An introduction emphasizing chemical equilibria in natural waters. New York: Wiley-Interscience, 1970. 583 p.

Turekian, K. K. Oceans. Newark: Prentice-Hall, 1968. 120 p.

U.S. Atomic Energy Commission. Precipitation scavenging. Washington: U.S. Atomic Energy Commission, 1970. 499 p.

Organic
Geochemistry

Organic geochemistry considers the composition, molecular state, spatial distribution and evolution of organic molecules, wherever they are found in the cosmos. Organic geochemistry can be distinguished from biochemistry in that it usually does not study the processes within living organisms. It is interested in what goes into and out of them, and in their bulk compositions, but not in what goes on inside. The naturally occurring organic compounds in the crust of the earth are, for the most part, the remains of once-living organisms. Complex carbon-containing molecules, probably formed from simple precursors by abiological processes, occur in extraterrestrial matter.

Organic geochemistry is interdisciplinary, combining aspects of geology, chemistry and biology. Because it is relatively new, much effort has been spent in defining the range of naturally occurring organic matter. Specific, specialized areas of research are emerging. Many organic geochemists are employed by the petroleum industry, wherein much research, both applied and basic, is being undertaken.

Organic molecules contain carbon in chemical combination with hydrogen, oxygen, nitrogen, and sulfur, as well as with many other elements present in trace amounts. Although carbon is the fourth most abundant element in the cosmos, it is only eighteenth in the earth's crust. Although it is a minor terrestrial element, a separate discipline has arisen to deal with the complexities of its compounds in geological and cosmological materials.

An estimate of the distribution of reduced and oxidized carbon is shown in Table 3. The numbers are provisional and their accuracy is uneven; they depend on the total masses of the individual reservoirs and the concentrations of both reduced and oxidized carbon in each reservoir. The errors in these parameters may be large, especially for igneous and metamorphic rocks.

Shales are the major reservoir of organic carbon on earth, but little is known about the variety, structure, or composition of their disseminated organic molecules. Petroleum in reservoirs, however, with a mass only 0.01 percent of shale organics, has been well studied because of its economic importance. The nature of the organic matter in the biosphere is well known, although it is a relatively small reservoir in the total carbon cycle. The nature of the organic carbon dissolved in the hydrosphere is not well-known because of difficulties in sampling and analyzing it. It is, however, extremely important be-

TABLE 3 Geochemical Inventory of Carbon in the Earth's Crust[a] (in Units of 10^{18} Grams)

	Graphitic or Organic Carbon	Carbon as Carbon Dioxide or Carbonates
Volcanic rocks	2,200	2,700
Granitic rocks	1,300	1,000
Metamorphic rocks	3,500	2,600
Sedimentary rocks		
Claystones and shales	8,900	9,300
Carbonates	1,800	51,000
Sandstones	1,300	3,900
Petroleum		
Nonreservoir	200	
Reservoir	1	
Coal	15	
Dissolved in hydrosphere	0.7	38.3
Atmosphere		0.64
Biosphere (land and water)	0.3	
TOTAL	~19,000	~71,000

[a] Hunt, J. M. Distribution of carbon in crust of earth. Bull. Am. Assoc. Petrol. Geol. *56*: 2273 (1972).

cause it regulates much of the biological activity in the oceans and because it represents organic matter en route to depositional sites in sediments.

Carbon is unique in its ability to form large molecules with bonds between atoms of the same kind. The complexity of the 3-dimensional architecture of its molecules is unequalled by that of any other element. There are about 10^2–10^4 inorganic minerals, depending on the definition of a mineral, but as many as 10^5 different kinds of molecules occur in living cells. These in turn are transformed into many more kinds during their residence in sediments. The variety and complexity of the molecules require sophisticated experimental work to identify them, but once they are identified, they increase the information available from an organic assemblage in a rock. The detailed information is particularly useful in deciphering the nature of precursor compounds and pathways of organic transformations. A start has barely been made on extracting and interpreting this information.

Organic geochemistry must deal with reactive molecules that are not thermodynamically stable substances and are not in thermodynamic equilibrium with their surroundings. Many of the reactions tending toward equilibrium are extremely slow; unstable complex organic molecules even occur in rocks of Precambrian age.

HISTORICAL PERSPECTIVE

Deposits of organic matter in the earth have been studied and described for a long time. The present goal of organic geochemistry is to understand this material at the molecular level and to study the processes it has undergone. Only recently have tools adequate to accomplish this task been available.

The American Petroleum Institute Project 6, started at the National Bureau of Standards in 1927, has been a landmark in the progress of organic geochemistry. Precision distillation columns and other methods of separation permitted a systematic study of the molecular constitution of a typical petroleum. The low-boiling constituents were separated, identified, and their weight fractions determined. The attack on more complex, higher-boiling-point compounds had to await separation methods with higher resolving power. One of the many significant results of the excellent work done on the project was the discovery that although petroleum is a complex mixture, it is composed of relatively few of the theoretically possible molecules. Many of these could be isolated and identified.

A second landmark study was Treibs' isolation and identification of metalloporphyrins in shales and petroleums in the 1930's. The compounds were obviously transformation products of pigments from living organisms. Although the mixtures of porphyrins are more complex than Treibs originally thought, the work convincingly demonstrated biological-marker molecules and signaled the end of serious discussion of the abiological origin of petroleum.

Rapid advancement came in the early 1950's with the introduction of chromatographic methods of separation and sensitive spectroscopic methods of structure determination. The course of future work was set, and an identifiable discipline was founded.

An ever-increasing amount of basic knowledge has been contributed by the petroleum industry in the past 25 years. Also, a period of rapid growth occurred recently when the principles of organic geochemistry were applied in the space program for the purpose of developing concepts and techniques for a search for extraterrestrial life. Many of the procedures in use today are the result of development work done within the space program.

CONDUCT OF CURRENT RESEARCH, SOME RECENT ACHIEVEMENTS, AND DIRECTIONS FOR THE FUTURE

Because organic geochemistry is a relatively young science, its first phases have been largely descriptive, with emphasis on analytical results. The rapidly changing situation has moved the stress to understanding the geological processes in which the organic compounds have been involved.

A generalized outline of the transformation of organic matter in geological settings is given in Table 4. Not all organic geochemists will agree on the fidelity with which it represents the actual processes that occur in nature; the sequence of events outlined suffers from being too general and qualitative. Coal geochemists have long used a somewhat similar semiquantitative description of the coalification process. Table 4, however, summarizes a number of facts, provides a convenient framework for discussing the current status of the field, and points to future developments.

The molecular constitution of living matter (biopolymers) is fairly well-known and provides the ingredients that enter geochemical sys-

TABLE 4 Geochemical Alteration of Organic Matter with Increase of Time and Temperature

Molecular Types (Processes)	Examples
I. Biopolymers (hydrolysis, microbial degradation)	carbohydrates, proteins, lipids, lignin, nucleic acids
II. Biomonomers (sedimentation, condensation, polymerization)	sugars, amino acids, fatty acids and alcohols, sterols, terpenoids, hydrocarbons, phenols, quinones, aromatic acids, purines, pyrimidines, and pigments
III. Geopolymers (diagenesis, thermal reactions, cracking, disproportionation)	"humic acid" complex, polymerized lipids, resins, brown coal
IV. Geomonomers plus Higher Polymers (metamorphism)	petroleum, gas, bitumen, kerogen, and bituminous coal
V. End Products	graphite, anthracite, methane, carbon dioxide

tems. There is good evidence that a number of the biosynthetic pathways used by organisms today have been used during much of the geologic past. No doubt some minor components of organisms are unknown, but the major constituents are well-defined. On the death of an organism and disaggregation of its cells, the components are subject to both aerobic bacterial metabolism and inorganic oxidation processes. This is the fate of most organic matter. Indeed, it is remarkable that any organic substances remain to be deposited in sediments. It has been estimated that photosynthesis throughout geological time has fixed an amount of carbon about equal to the mass of the entire earth. Because such a small fraction remains in rocks, it is surprising that there were not long stretches of time during which no organic material was stored. Oddly enough, the percentage of organic material in rocks is nearly independent of geologic time.

In restricted environments that are low in oxygen content, some material is preserved and slowly degraded into biomonomers. This stage must be brief in terms of geologic time, because most biomonomers are reactive and either degrade spontaneously into simple molecules or react with other compounds. The third stage, formation in sediments of complex polymers, often called geopolymers, is achieved early in the history of organic materials. The bulk of organic carbon in a soil or recent sediment consists of dark, insoluble polymeric material. It has an elemental composition not greatly different from many living organisms, but its molecular composition has been changed considerably.

With time, buried in sedimentary systems, organic matter begins to undergo transformations, processes known as diagenesis. The rates of these processes are controlled by temperature and by the chemical conditions of the sedimentary environment. During diagenesis, the complex polymers condense to higher-molecular-weight substances and undergo disproportionation; increasingly stable fragments of low molecular weight split off. With increased time and elevated temperature during metamorphism, organic carbon is finally transformed into the end products, graphite, anthracite, methane, or carbon dioxide.

At each of the steps in Table 4, there is the possibility of oxidation by biological or inorganic processes and recycling of the carbon as carbon dioxide. The separation of the stages is not always clear-cut. It is improbable that one could find a situation where all steps of Table 4 could be traced through a given sedimentary rock assemblage; however, some of the long continuous cores taken by the Deep Sea Drilling Project of the National Science Foundation permit examination of early stages. In other instances, for example, where an ancient sedimentary rock sequence has been intruded by molten rock, the last (metamorphic) stages have been studied.

Molecular Studies

An examination of the published literature in light of Table 4 shows clearly that knowledge of the geochemistry of certain monomers, among them hydrocarbons, fatty acids, porphyrins and amino acids, has reached a higher level of sophistication than other parts of the field. In some instances, for example, in isoprenoid lipids and amino acids, even the detailed molecular architecture (stereochemistry) has been explained, and details of source material and of reaction pathways have been obtained by use of highly refined analytical techniques. The reason hydrocarbons, fatty acids, porphyrins, and amino acids have been studied extensively is obvious: These classes of compounds are readily extracted and isolated. The instruments, such as spectrophotometers, fluorometers, gas chromatographs, liquid chromatographs, and mass spectrometers required to identify these compounds are available; research in organic geochemistry is determined largely by existing instrumentation. The study of these four classes of compounds is being taken rapidly from the context of pure organic analytical chemistry to a blend with sedimentary geology, chemical taxonomy, and paleobiogeochemistry.

Other classes of organic compounds have received less attention; it is time to extend analyses to such classes of compounds as sugars, nucleic acid bases, alcohols, phenols, and pigments excluding porphyrins now that modern analytical methods have the required sensitivity. The chemical stability of these compounds in terms of the geological time scale is not well-known. Some may disappear rapidly. The molecules are sufficiently complex, however, and exist in such wide variety, that information on at least the first steps of transformation in geological environments should be obtained. The ultimate goal of this work is to understand the occurrence, relations, and fate of all significant organic materials in the geologic record.

Geochronology and Paleothermometry

Techniques have been developed whereby the relative amounts of optical isomers of amino acids can be measured. With these techniques, it is now possible to determine the extent of racemization (decline in optical activity) of amino-acid enantiomers (mirror images), and the rate of racemization has been shown to be dependent to a large extent on temperature. Determinations of this kind are potentially useful in establishing a new method of determining the ages and thermal histories of sedimentary deposits. In carefully controlled environmental situations where the temperature is known, the age can be determined; where the age is known, the average temperature during that time period can be calculated. As a geochronologic tool, the method appears to be useful for determining ages ranging from zero to about 12 million years, and possibly even older. The time range for the method, therefore, overlaps and extends the radiocarbon method (0–40,000 years) and requires only micrograms of amino acids, whereas the radiocarbon method requires about a gram of carbon.

Paleontology

Through the stages of transformation of organic matter shown in Table 4, there is a diminishing record of the original biopolymeric compounds. Some molecular information is lost quickly, but some is retained for a long time. The situation is analogous to the irregularities in preservation of the fossil record and, indeed, certain organic compounds are described as "chemical fossils" or "biological markers." These biopolymeric molecules that have persisted since the Precambrian have received much attention.

It is logical to attempt a correlation of morphological fossils with chemical fossils. In rocks of young age, this correlation will be most evident, and contamination will probably be minimal. With knowledge of the chemical fossils in young samples, extrapolations of results to older rocks can be made with more confidence. For example, the amino acids in shells of the mollusc *Mercenaria mercenaria* have been traced from living specimens back to Upper Miocene fossils some 11 million or more years old. Biosynthetic pathways such as those of today have been demonstrated, and the potential for discovering factors affecting biochemical as well as morphological evolution has been shown.

At the other end of the time scale, in the Precambrian era, morphological fossils are scarce, and for studies concerned with early life on earth, other kinds of evidence must be sought. Well-preserved sedimentary rocks, 1200 million years old, are known to contain abundant complex organic molecules. Some of them are typical biological markers, such as normal and isoprenoid hydrocarbons. Although the exact pathways of formation of these compounds are not known, they were undoubtedly produced from the lipid fraction of living organisms. It is not known at present how far back in time one can safely interpret the organic geochemical record. Carbon-containing rocks older than 2000 million years and recognizably sedimentary in origin with a low-temperature postdepositional history are rare. Only a few complex organic molecules appear to have the intrinsic stability to persist for periods exceeding 2000 or 3000 million years, even in the most favorable environments. Continued geochemical studies of biological markers in rocks of all ages will provide a record of paleobiochemistry and will contribute to deeper understanding of paleontology.

Complex Organic Polymers

Although many organic monomers are easily extracted and identified, they often constitute only a small fraction of the organic carbon in rocks. Characterization of the more abundant, intractable complex organic polymers has proceeded much more slowly. Yet the origin of this complex polymeric material, which constitutes the bulk of carbon in most natural samples, may be the most important and most difficult problem in both organic geochemistry and organic cosmochemistry. In fact, in many sediments and meteorites this polymer could be considered a major element because it may constitute at least one percent of the rock. The study of the structure and constitution of this polymer has proved to be difficult and, so far, unrewarding. The application of some of the classical degradation procedures of organic chemistry has been only partly successful. Perhaps new techniques are required, but imaginative new approaches with old techniques should also be tried. Models for the origin of organic polymers are few. The reactants and reactions that lead to the first steps of polymer formation in sediments are obscure; a great many possibilities could be important in the process. An approach from synthetic organic chemistry may be valuable if applied with modern methods. Only with more complete understanding of this organic polymeric material will a complete picture of the carbon cycle be developed.

Laboratory Simulations

Many geochemical processes can be simulated in the laboratory, where elevated temperatures substitute for the long time intervals of earth history. Thus reactions that occur slowly in sediments can be reproduced rapidly in the laboratory. Temperature and pressure conditions have been applied to organic materials interpreted as precursors to other materials for varying periods of time. The resulting products are then studied in terms of the precursors and the imposed conditions. The fate of organic monomers and polymers can be conveniently studied and correlated with actual occurrences in natural systems. These kinds of laboratory studies will be useful in the future in providing tests for hypotheses generated on the basis of observations of naturally occurring samples.

Environmental Organic Geochemistry

The techniques and methods of organic geochemistry are directly applicable to environmental studies. The baseline to which pollution is contributing must be ascertained and this is the direction of some current research. The fate of synthetic materials such as pesticides and detergents should be investigated in more detail. The effects and fate of large concentrations of organic material, such as oil and sewage, resulting from man's activities, should be subjects of direct interest to organic geochemists. Recent experiments have begun to utilize radiolabeled organic chemicals to try to understand the rapid reactions undergone by organic compounds. In these studies, radio-labeled lipids have been added to recent sediment, both *in situ* and in the laboratory. The fate of the label has been determined, thus permitting a direct measure of reaction pathways. The use of radiolabels in many different kinds of molecule, including pesticides and detergents, is clearly indicated to gain information about processes that take place early in the history of sediments.

Microbiology

Microbial activity plays an important role in geochemistry, but the extent of this role is not clearly known. It is known, however, that most carbon of organic compounds of the biosphere is completely metabolized to carbon dioxide by microorganisms in soils and sediments. The extent to which some compounds are only partly transformed is uncertain. Nevertheless, many reactions of organic materials in the water column and in early sediments are greatly influenced by microorganisms. Microorganisms are important in the alteration of petroleum deposits exposed to fresh water. They are active in some ore-forming processes and are important as part of the weathering process of rocks. Although geochemists tend to overlook the activities of microorganisms, it is becoming increasingly apparent that the results of their activities must be ascertained before complete knowledge of some geochemical processes is obtained.

Stable Isotope Geochemistry

The biogeochemistry of the stable isotopes of carbon and sulfur is well-developed. Studies of variations of abundances of the stable isotopes of carbon, ^{12}C and ^{13}C, have become useful adjuncts in a number of areas of geochemical research. Carbon isotopes are used as tracers for source rock–petroleum correlations. They have been applied to a wide group of problems ranging from studies of sources of organic detritus in recent sediments to studies relating to photosynthesis during the Precambrian. Most of the measurements have been made on mixtures of organic compounds, but current trends point to work on single, pure compounds or even to individual carbon atoms within compounds. There is a wealth of information regarding isotopic fractionation processes contained in the detailed distribution of stable carbon isotopes of individual molecules. The relative importance of the steps in a complex fractionation process can be assessed. The fractionation of the stable isotopes of sulfur, ^{34}S and ^{32}S, associated with sedimentary organic matter is now an important part of research on the exogenic cycle and further study of this phenomenon will undoubtedly increase its potential.

Comparatively little is known about the biogeochemistry of hydrogen- and nitrogen isotopes, and this area of research is ready to be exploited. The concentration of deuterium in water has a well-known geographic distribution. An isotopic tracer for land-derived organic matter exists in organic molecules, which permits this organic matter to be distinguished from that synthesized in the oceans. A major problem is to determine whether isotopic exchange occurs at a rate that would cause the tracer to be lost. The lipid fraction of soils and sediments is the most obvious fraction to study. Further experimentation with the nitrogen isotopic effects in microbiological cycles is needed. An important potential application of stable nitrogen isotopic techniques is determination of the contribution of nitrate fertilizers to water supplies. More must be known about the biogeochemistry of nitrogen before this area of research can be fully exploited.

Isotopic abundance patterns in lunar samples and meteorites appear to be different from those observed in terrestrial substances. Studies of isotopic fractionation in the samples may lead to understanding of mechanisms involved in extraterrestrial organic reactions.

Organic–Inorganic Interactions

The role of organic–inorganic associations in molecular transformations of organic molecules and in transportation and adsorption phenomena is not well-understood. It is known, for example, that much of the organic matter in recent marine sediments has been transported to the site of deposition by means of clays. A considerable body of information about clay-organic associations has been gathered from "clean-laboratory" studies. Future studies should include experiments with the complex mixtures found in natural environments. Such experiments will apply directly to geochemical studies of the dissolved and particulate material of the hydrosphere. Studies on a molecular level depend on proper sampling methods. Simple contamination-free methods of isolating and concentrating organic materials from water are needed.

Because organic matter constitutes a small fraction of most rocks, there is controversy over the role played by mineral surfaces during diagenesis. If weak physical absorption is predominant, surface effects may be unimportant. On the other hand, mineral surface-organic interactions may be strong, and catalytic effects could control the rates of change of organic materials. Laboratory and field studies are needed.

Calcification

Apparently organic matter is important in the crystallization of certain inorganic minerals. For example, inorganic processes such as calcification are heavily influenced by the presence or absence of organic materials. These substances either promote or inhibit calcification and consequently must be investigated if this process is to be completely understood. The role of organic matter in what initially appeared to be inorganic processes is becoming increasingly evident. Sensitive techniques are now available for organic analyses, and identification of the organic molecules is being made. Additional efforts in these areas of mutual interest to organic and inorganic geochemists are indicated.

Economic Deposits

Organic geochemistry has played an obvious role in the petroleum industry. Through geochemical studies, the compositions of oils and natural gases have become known, and working hypotheses concerning the origin of petroleum have been generated. Progress has been made by applying considerations of organic reaction mechanisms to the transformation of organic chemicals during the petroleum-forming processes. Details about natural thermal cracking and disproportionation of protopetroleum have been learned. Observations of precursor–product relations in field samples have proved to be useful; for example, studies of coexisting fatty acids and saturated hydrocarbons in shales have been made. Although there is still some disagreement on the extent to which free fatty acids are decarboxylated to give hydrocarbons, the studies have set limits to speculations on the origin of petroleum. Geochemistry has aided in petroleum prospecting, particularly in identification of potential source rocks. Further studies should reveal more intimate details of the process of petroleum formation and improve the efficiency of petroleum exploration.

Although organic matter is chiefly associated with sedimentary rocks, its association with some hydrothermal fluids, ore veins, and fluid inclusions is known; however, the significance, if any, of the organic material in ore-forming processes is not known. In a few of the instances studied, the organic matter appears to be of biological origin, originally deposited in sediments. Somehow it has been transported in mineral-forming solutions. Because of the economic importance of ores, studies of the association of organic matter with ore deposits may be useful.

Extraterrestrial Organic Material

The discussion, so far, has been concerned mainly with organic chemicals from living organisms that have been deposited in geologic systems. Parallel to this field, studies have been made of organic molecules produced abiotically in experiments designed to simulate the primitive atmosphere of the earth. Since the early 1950's, much evidence has been produced to show that complex mixtures of organic molecules can be generated easily from simple carbon compounds by nonbiological means. It is agreed that this process is not of quantitative importance on earth at present, and has left little or no record in ancient rocks, even though it may have been critical in the origin of life on earth, which precedes the terrestrial sedimentary record. Also, similar abiological processes are a possible source of extraterrestrial organic matter. Recent studies of carbonaceous meteorites have tended to support this hypothesis. Early studies of carbonaceous meteorites were ambiguous because of the unknown conditions they experienced during their fall to earth and during subsequent storage and handling. The small amounts of monomeric organic compounds that have been studied have a large and variable component of terrestrial contamination. The polymeric organic matter that forms the bulk of organic carbon has not been well-characterized. The Murchison meteorite (Murchison, Australia, September 28, 1969) is an example of a large fall that was collected soon after it fell and had minimum terrestrial contamination when it was initially subjected to chemical examination. A suite of amino acids and of hydrocarbons was found that has been interpreted to be of extraterrestrial origin.

Further discussion of extraterrestrial organic material is found in the chapter, "Extraterrestrial Geochemistry."

Origin of Life

Evidence related to the origin of life may be contained in meteorites and in the early Precambrian rocks of earth. The organic compounds in meteorites may be interpreted as precursors of life; whereas much of the organic material in Precambrian sediments appears to have been generated after life began. Nevertheless, if any record of the origin of life exists, it must be locked in the Precambrian rocks that span 75 percent of recorded geologic time. It is unlikely that individual monomers are present in rocks more than 3-billion-years old. Studies of the properties of the bulk carbon may provide clues to ancient atmospheric conditions. Comprehensive studies of elemental and isotopic abundances of carbon and other organogenic elements may show important trends that will yield answers to fundamental questions about life on earth.

Major Future Goals

Organic geochemistry has at least two major goals for the future—the refinement of techniques, and the application of resulting improvements to the solution of important geological and cosmological problems. Of these two, the second is by far the more important. To achieve it, a multidirectional approach to problems of broad geologic scope should be encouraged, in which organic geochemical, inorganic geochemical, and geological information is collected and interrelated whenever possible.

A trend toward this kind of study is already evident. Recent studies of the Mackenzie River system, Saanich Inlet (British Columbia), the Dead Sea, and the Red Sea have engaged a number of investigators who have made a variety of different measurements and have attempted to correlate their findings to understand processes within these important and interesting geological and geographical regions. Extensive geochemical studies of the continental shelves represent areas for this kind of large-scale approach. Studies of the terrestrial and marine environments have been subject to artificial separation; the most fruitful field of all may be study of the transition from one to the other. It is in the estuaries and on the continental margins and shelves that the effects of man are so strongly felt. Team studies will bring together marine geochemistry and terrestrial geochemistry to solve problems in regions of direct importance to man. Some of the organic geochemical topics that could be investigated have been discussed above.

SELECTED READINGS IN ORGANIC GEOCHEMISTRY

Degens, E. T. Geochemistry of sediments—a brief survey. Prentice-Hall, 1965. 342 p.

Eglinton, G., and M. T. J. Murphy (eds.). Organic geochemistry—methods and results. New York: Springer-Verlag, 1969. 828 p.

Gaertner, H. R., and H. Wehner (eds.). Advances in organic geochemistry, 1971. Braunschweig: Pergamon Press, 1972. 736 p.

Hood, D. W. (ed.). Organic matter in natural waters. College: University of Alaska, Institute of Marine Science Occasional Publication No. 1, 1970. 625 p.

Hood, D. W. (ed.). Impingement of man on the oceans. New York: Wiley-Interscience, 1971. 738 p.

Massachusetts Institute of Technology. Man's impact on the global environment: Report of a study of critical environmental problems. Cambridge: M.I.T. Press, 1970. 319 p.

Singer, S. F. (ed.). Global effects of environmental pollution. New York: Springer-Verlag; Holland: D. Reidel Publishing Co., 1970. 218 p.

Wedepohl, K. H. (ed.). Handbook of geochemistry. New York: Springer-Verlag, 1969.

Geochemistry
and Environmental
Concerns

As stated earlier, geochemistry is concerned with chemical cycles in nature, such as the circulation of materials from ocean to atmosphere to land to streams to ocean and back again to land. The return to land may be by burial and uplift or by escape from the sea surface. In this sense the earth has a natural metabolism: For hundreds of millions of years it has been producing and destroying innumerable natural compounds in these cycles. A number of pollutants can be considered simply as increased injections of natural compounds into these cycles—additions to the atmosphere of gases containing sulfur, nitrogen, carbon dioxide, carbon monoxide, and mercury and its compounds; additions to groundwaters and streams of toxic minor elements such as lead, zinc, arsenic, and selenium; additions to lakes and the ocean of nutrients, animal wastes, petroleum, and radioactive nuclides.

Other pollutants are new to the earth. Important among these are such synthetic organic chemicals as the halogenated hydrocarbons used for agricultural, public health, and industrial purposes. These volatile pollutants are found in the atmosphere as gases, in seawater, in all forms of marine organisms, in groundwater—in short, throughout the exogenic or surface environment.

Synthetic chemicals often follow pathways and undergo reactions in their cycles similar to those of natural materials. Thus, an understanding of natural cycles is necessary for assessing the effects of synthetics on the environment. Geochemists can have a substantial impact on environmental problems in assessing the long-term effects of man's interference with natural chemical cycles. This interference and its future effects can only be understood from knowledge of preman cycles and the complex feedbacks that have maintained life on the earth for over 3 billion years.

In a complementary way, investigations of the behavior of man-mobilized species in the environment have increased our knowledge of the composition and reactions of natural systems. Injections of radioactive materials into the atmosphere and oceans through nuclear detonations have provided a host of radioactive tracers for natural processes. For example, the differences in behavior of oceanic strontium, cesium, and the rare earths have been well-delineated. The discovery of the production of methyl mercury and dimethyl sulfide compounds by lower organisms during studies of lacustrine and atmospheric pollution problems, has helped in the understanding of the natural cycles of mercury and sulfur on the earth. Environmental pollution studies will result in other advances in basic geochemistry.

The following sections will summarize some natural chemical cycles and feedbacks and man's interference with these cycles. In particular, gaps in current knowledge will be highlighted. This approach should help in understanding the interrelations between geochemistry and environmental concerns, and may delimit some of the areas in which geochemists can contribute to alleviation of environmental problems.

CYCLES OF CARBON, NITROGEN, AND PHOSPHORUS

Synopsis of Cycles

The organic cycles of carbon, nitrogen, and phosphorus apparently have been relatively unchanged for the past 600 million years. This constancy is best documented for organic carbon. Organic matter in rocks is oxidized to such species as CO_2 and NO_3^-, or nitrogen gas (N_2), during weathering, and phosphorus is released as phosphates by oxidation and hydrolysis. Nitrate and phosphate enter stream waters. Weathering of organic matter removes molecular oxygen from the atmosphere and increases carbon dioxide. The O_2 is restored by photosynthesis and a balance is maintained by deposition of organic matter, chiefly in the ocean. The cycle is eventually completed as the deposited organic matter is uplifted and on exposure again subjected to oxidation. Because the average carbon/nitrogen/ phosphorus ratio of organic matter in sedimentary rocks differs from that in living matter, selective return of carbon, nitrogen, and phosphorus to the ocean during the early stages of burial of organic matter is required to maintain the mass balance.

The exchange relations among the various reservoirs in the carbon cycle, and probably the nitrogen and phosphorus cycles, have been delicately balanced for the last 600 million years. Evidence for this conclusion stems from observations that no long-term trends are evident in the δ ^{13}C composition of limestones or in the organic carbon content and carbon/nitrogen ratios of sedimentary rocks for the past 600 million years. The fluxes of organic carbon, nitrogen, and phosphorus into sediments have not varied greatly during this time. For carbon, the flux represents the residual from photosynthesis minus decay and respiration and is only about 0.05 percent of the total carbon fixed annually. This delicately balanced cycle has been seriously influenced by man since the beginning of the Industrial Revolution.

Man's Interferences

The carbon dioxide in the atmosphere is controlled by the demands and releases during natural weathering, by respiration and decay processes, by mineral deposition in the ocean, and by the burning of fossil fuels. The burning of fossil fuels has increased the rate of addition of carbon dioxide to the atmosphere; man's contribution is about 10 percent of the total carbon fixed by photosynthesis each year. The pre-1959 carbon dioxide level of the atmosphere was about 313 ppm; 10 years later the level was 321 ppm, and by the year 2000 it could rise to 375–400 ppm.

The effects of higher atmospheric carbon dioxide concentration on the earth's climate are still controversial. Atmospheric warming owing to an increased "greenhouse effect" may occur. Before definitive predictions can be made, however, the effect on climate of variations in parameters such as atmospheric moisture and cloud coverage, changes in the extent of arctic sea ice, and in the amount of atmospheric particulate matter, needs to be evaluated.

The sinks for carbon dioxide are photosynthesis, the terrestrial biosphere, and accumulation in atmosphere and ocean. Indeed, only about 30–40 percent of the carbon dioxide introduced into the atmosphere by man's use of fossil fuels still remains there. The rest has either dissolved in the ocean or increased the mass of the terrestrial biosphere; one gap in our knowledge is the relative importance of these two reservoirs as sinks.

Both carbon monoxide and methane are added to the atmosphere by fossil-fuel burning. The sources of natural carbon monoxide additions to the atmosphere have been investigated as a result of pollution problems, and it appears that only a few percent of the normal atmospheric burden of carbon monoxide is produced by man. The level in the atmosphere is determined by the photochemical decomposition of formaldehyde produced from the oxidation of methane. The sources of methane are probably the anoxic zones of the earth's surface: swamps, fjords, and paddy soils.

The addition of nitrogen and phosphorus compounds to the earth's surface owing to man's activities often causes an increase in the rate of eutrophication of water bodies on a local as well as on a regional scale. Nitrogen is dispersed on the earth's surface in fertilizers, sewage, animal wastes, and in NH_3, NO, and NO_2 produced by combustion of coal and petroleum. Nitrogen fixation today exceeds return of nitrogen by about 10 percent because of cultivation of legumes and industrial fixation to make fertilizers. Fixation of atmospheric nitrogen in gasoline-powered piston engines is also an important factor. It is the excess fixed nitrogen that leads to increased rates of eutrophication of water bodies.

Phosphorus is added by man to the natural phosphorus cycle principally from fertilizers and detergents. Man now mines and uses annually about 10 times more phosphorus than was carried in solution to the ocean by rivers before man arrived.

Some Research Needs

To predict the effects pollutants will have on the cycles of carbon, nitrogen, and phosphorus, it is necessary to fill some major gaps in knowledge, particularly with respect to the global fluxes of pollutants and to prediction of their future impact on the earth's surface environment.

• There is a need to construct both steady-state and time-dependent models for these cycles. The time-dependent models will be particularly revealing, because additions of pollutants are time-dependent variables. The usefulness of these models will depend to a significant degree on the knowledge of processes affecting carbon, nitrogen, and phosphorus behavior at the earth's surface.

•   To understand how an increased rate of addition of carbon dioxide will affect the earth's surface environment, better values will be necessary for the present mass of carbon in various reservoirs, particularly for factors governing transfers among reservoirs—present values are semiquantitative. A good model of the system will necessitate values not only for carbon but also for other elements, such as nitrogen and phosphorus, involved in the fixation of carbon in organic matter and in its release by decomposition and decay.

•   It is unlikely that the recent exponential increase of carbon dioxide in the atmosphere will continue indefinitely. Nevertheless, predictions of any rate of increase will require more knowledge of the controls of the preman cycles of carbon, nitrogen, and phosphorus. Although these cycles apparently have been remarkably constant, there is evidence in the sedimentary record based on the mineral phases present that suggests that changes have occurred in, for example, the carbon dioxide content of the atmosphere during geologic time. The magnitude of these changes and the rates at which they were dissipated need to be evaluated before predictions of the future impact of carbon dioxide addition to the atmosphere can be evaluated.

•   There are other gaps in knowledge of the carbon, nitrogen, and phosphorus cycles that need to be filled. For example, the distribution of phosphorus among inorganic phases and organic matter in sedimentary rocks is unknown. The amounts of carbon, nitrogen, and phosphorus carried to the oceans in the dissolved and particulate organic loads of streams are poorly established. Soil reactions involving organic matter deserve more attention, so that the fluxes of carbon, nitrogen, and phosphorus compounds into groundwaters, streams, atmosphere, and ocean can be determined. During shallow burial of organic matter in sediments, compounds of carbon, nitrogen and phosphorus dissolve in interstitial waters and diffuse upward. Fluxes of these compounds into overlying waters need to be measured, as well as the complex organic reactions that lead to release of these compounds. Lack of knowledge of the atmospheric loop of the nitrogen cycle is discussed in the chapter on exogenic geochemistry.

•   Little is known about the natural production and global cycle of methane. Geochemists could contribute significantly to understanding of the cycle by evaluation of the reactions leading to production of methane in soils and anoxic sediments and determination of the methane flux out of sediments.

• On a local basis, geochemists could contribute to an understanding of lake and river eutrophication by studying the sources, fluxes, and sinks of the involved organic matter. These and other investigations involving the cycles of carbon, nitrogen, and phosphorus would benefit greatly from interactions among organic and inorganic geochemists and biologists.

SULFUR CYCLE

Synopsis of Cycle

About two thirds of the sulfur in sedimentary rocks occurs as reduced sulfur in pyrite; the rest is found mostly as calcium sulfate. Under steady-state conditions, the weathering of calcium sulfate would release Ca^{++} and $SO_4^=$ to streams, and a mass of $CaSO_4$, minus that recycled through the atmosphere, would be deposited in the ocean. On the other hand, oxygen is required for the weathering of pyrite, creating a drain of O_2 on the atmosphere. The oxidized sulfur and iron are transported to the ocean where, in anoxic regions they react with organic matter to again form pyrite (FeS_2). Carbon dioxide is released and through photosynthesis can return O_2 to the atmosphere.

Sulfur isotopic studies of evaporites suggest that the ratio of oxidized sulfur to reduced sulfur ($^{34}S/^{32}S$) in sedimentary rocks has varied with time. This observation implies a major feedback mechanism involving the transfer of calcium from the calcium carbonate to the calcium sulfate reservoir and return of atmospheric oxygen by photosynthesis of released carbon dioxide.

Because of the lack of extensive deposition of calcium sulfate today, it is possible that the sulfur cycle is out of balance and oxidized sulfur is accumulating in the oceans as $SO_4^=$.

Natural sources of sulfur in the atmosphere are hydrogen sulfide generated by the decay of vegetation and bacterial reduction of sulfate [$(CH_3)_2 S$] produced metabolically by plants, $SO_4^=$ produced by oxidation of SO_2, sulfate particles derived from sea aerosols, and volcanic gases containing sulfur.

Sulfur is added to the atmosphere by man's activities principally as SO_2 through burning of fossil fuels. Some sulfate enters streams directly from the use of sulfur-bearing fertilizers. The magnitude of man's additions of sulfur to the natural cycle appears to be comparable to that cycling naturally. Current assessment of the effects of man's contribution indicates that although it has severe local effects, there has been little disturbance of the natural cycle, in the sense that cessation of man's additions would cause a quick return to preman conditions and that there has been little global effect by man on the sulfur content of the global atmosphere.

Some Research Needs

Knowledge of the natural chemical cycle of sulfur is necessary before we can assess the anthropogenic contribution and its effects. Time-dependent mathematical models of the sulfur isotope composition of calcium sulfate deposits can help in predicting what can happen to the sulfur cycle when it is perturbed by an event of short duration but large magnitude; in this case, the addition of pollutant sulfur.

Organic reactions involving sulfur in soils constitute a fertile field of research. A balanced sulfur cycle would be indicated if it could be shown that a significant quantity of H_2S is produced in soil zones from the decay of organic matter. Furthermore, it would be helpful to understand in greater detail the reactions involving sulfur-bearing organic matter in various environments.

Studies of the stable isotopes of sulfur in soil zones, as well as in river waters, may help in distinguishing natural from pollutant sources of sulfur. Also, additional studies of sulfur isotopes in sedimentary materials through time are necessary to understand the preman sulfur cycle and to predict the impact of pollutant sulfur on the environment.

HEAVY METALS

Toxic heavy metals, including lead, copper, chromium, mercury, cadmium, selenium, and zinc are being released to the surface environment at an increasing rate by mining and industrial use. Annual mining production of many of these metals already exceeds preman transport to the ocean by streams.

Little is known about the preman and present exogenic cycles of these elements. Reservoir magnitudes and fluxes among reservoirs of some of these elements are unknown; for others, only approximations can be made. The chemical, biochemical, and geochemical mechanisms for removal of most trace metals from the oceans are not known and need to be determined.

A brief discussion of the lead and mercury cycles will illustrate the problems involved and the possible avenues of geochemical research that can help to solve them.

Lead

Approximately 2.5×10^{13} metric tons of lead are present in crustal rocks. Chemical weathering of various minerals such as galena (PbS), and especially of lead contained in potassium feldspar, an abundant crustal mineral, introduces lead into river waters. The input of dissolved lead to the ocean from rivers is about one half that of lead in suspended solid particles. If the lead cycle is steady state, this soluble lead must be deposited in anoxic environments as sulfide and in oxygenated areas with clays and ferromanganese minerals and other unknown mineral sinks and eventually returned to the land surface via burial and uplift.

Emission of lead from internal combustion engines and otherwise from the burning of fossil fuels is about equal to the soluble lead transported to the ocean annually by streams. Thus a major pathway of transport of pollutant lead to the ocean is via the atmosphere. Lead concentrations in surface seawaters, and in coastal marine, lake, and glacier sediments have increased owing to the rain-out of lead from the atmosphere.

No model of the preman lead cycle is available. The construction of this model will demand a great deal more data concerning the distribution of lead among organic materials and about various inorganic phases in sediments and sedimentary rocks.

Mercury

Although data are insufficient to permit quantitative estimates, the principal natural source of mercury appears to be the degassing of the earth's crust. The increase in the mercury content of snows on the Greenland glacier in recent years suggests that man's alterations of the land surface may have increased the rate of the degassing. The weathering of mercury-bearing rocks, especially of shales, leads to a flux of mercury to the ocean via streams of perhaps an order of magnitude less than the rate of crustal degassing. As with lead, if the natural cycle of mercury is in steady state, then the amount of mercury added to the earth's surface environment by degassing and by chemical weathering must be compensated by a like amount sedimented in the ocean and returned to the land surface by means of burial and uplift.

Anthropogenic sources of mercury include fossil-fuel combustion, cement manufacture, and industrial and agricultural uses. The amounts of mercury from these sources are all a magnitude or more less than those from crustal degassing.

Some Research Needs

Although innumerable problems are concerned with the cycles of heavy metals, two in particular seem to merit attention.

The masses of heavy metals in various natural reservoirs and the mechanisms that permit fluxes between them are needed to obtain the preman cycles of these elements. The preman cycles will provide some baselines for assessing the long-term effects of present increased rates of addition of these elements to the earth's surface environment; again, mathematical modeling of these cyclic systems should be helpful.

Studies of the sources, pathways, and sinks (organic and inorganic) of heavy metals in natural environments need to be expanded. Soils and the waters and sediments of coastal regions are of primary interest; however, transport routes and partitioning of these elements among the organic and inorganic dissolved load of streams, as well as between the organic and inorganic fractions of the suspended materials in streams, are important areas of geochemical research. Aside from supplying necessary data, these studies are essential to any model used to assess the long-term consequences of recent increased rates of introduction of heavy metals into the environment.

PETROLEUM

Man's activities have greatly increased the rate at which petroleum and its by-products have been added to the earth's surface environment. Little attempt has been made to assess the sources, pathways, and sinks of individual chemical components of petroleum on a global basis. It is known, however, that man's activities inject petroleum into the environment at a rate perhaps 1000 times faster than the natural seepage rate of oil. Table 5 shows the magnitude of petroleum fluxes to the environment. Most of this petroleum eventually ends up in the ocean. The mass of hydrocarbons in petroleum introduced to the oceans annually by ship spillage and leakage alone is roughly equivalent to total hydrocarbon production by marine organisms.

Much still remains to be learned about the behavior of petroleum in natural environments, particularly in seawater. In general the light fractions of petroleum added to the ocean are known to evaporate, whereas other fractions are absorbed on particulate matter and sink; some dissolve, and others are oxidized by bacteria. Fuel-oil hydrocarbons, however, appear to be persistent; 2 years after a spill in Buzzard's Bay, Massachusetts, the hydrocarbons had not been totally degraded and were found in offshore sediments. The tarry residues of petroleum, in particular, persist for a long time.

Some evidence suggests that even the aromatic fraction of oil may degrade slowly enough to enter sediments in significant concentrations. Benzopyrene, a carcinogen, has been found in concentrations

TABLE 5 Annual Fluxes of Petroleum and its By-products (1969)[a]

Source	Flux (units of 10^6 metric tons/year)
1969 World oil production	1820
1969 Oil transport by tanker	1120
Anthropogenic injections into marine environment	3.1
Seepage from wells	0.5
Tanker operations	0.5
Other ship operations	0.5
Accidental spills	0.2
Deliberate dumping	0.5
Refinery operations	0.4
Industrial and automotive wastes	0.5
Torrey Canyon discharge (off Land's End, England)	0.1
Santa Barbara, Calif., blowout	0.003–0.1
Vaporization of petroleum products from continents into atmosphere	90
Natural seepage	0.1

[a] Adapted from Massachusetts Institute of Technology. Man's impact on the global environment: Report of a study of critical environmental problems. Cambridge: M.I.T. Press, 1970. 319 p.

up to 5 ppm in marine sediments. It apparently can be taken up in marine plankton.

Petroleum and its refined petrochemical products are complex mixtures of paraffins, aromatics, asphaltics, and a wide variety of other constituents. Study of the natural cycles of these constituents is to a large extent the domain of organic geochemistry. There are at least two major avenues of research to be explored to aid our understanding of the long-term effect of man's addition of petroleum and its by-products to the environment.

The sources, pathways, and sinks of petroleum in coastal environments need to be documented. Coastal regions are particularly useful in geochemical studies because they are regions of complex interactions among organic and inorganic materials. In addition to reactions in seawater there are heterogeneous reactions between suspended material and seawater and interstitial water and sediments.

An assessment needs to be made of the sources, pathways, and sinks of petroleum and petrochemical constituents on a global scale before man's activities further upset the global balance. If possible, mass balances should be made for preman petroleum constituents.

SYNTHETIC ORGANIC CHEMICALS

Thousands of synthetic organic compounds, entirely foreign to the natural system, enter the environment each year. Those of particular interest are produced in large quantity, are relatively stable, and known to be toxic. These compounds include, for example, the organic chemicals DDT, PCB (the polychlorinated biphenyls), freons, and dry-cleaning solvents. DDT is used to illustrate how our information about these compounds and the role of geochemistry can play a part in evaluating the impact of these chemicals on the surface environment.

The global production of DDT is at present about 0.1×10^6 metric tons per year, of which perhaps as much as 25 percent reaches the sea, principally by way of the atmosphere. The rest presumably accumulates in terrestrial environments. DDT is removed from the atmosphere by rain and dry fallout.

Recently the United States Environmental Protection Agency banned the use of DDT, except in cases of public health emergencies and in application to a limited number of crops. DDT is still being used in many other countries, however, and because the atmosphere is a major pathway of dispersal of DDT, the United States ban cannot control global dispersion.

DDT is a universal poison, moderately volatile and persistent. It may take 10 years or more to degrade half of the amount added in any given year. It is insoluble in water but moderately soluble in organic phases and particularly in fatty tissues. It is these characteristics that have led to its wide dispersal throughout the global environment and its reputation as a dangerous chemical. Increases in concentrations of DDT in successively higher trophic levels in the biosphere have been reported and have been cited as the cause of the population decline of various carnivorous birds. DDT concentrations in phytoplankton populations in Monterey Bay, California, are at least three times greater today than in 1955.

The monitoring of DDT and other synthetic organic chemicals on a local and global basis, is suggested by the 1970 National Academy of Sciences report by the Ocean Science Committee, *Marine Environmental Quality*. The reaction rates and chemical relations of DDT and its residues and of other organic chemicals may be an area of fundamental interest to geochemists, particularly because these rates and mechanisms are probably of a heterogeneous nature and involve sediment particles in terrestrial and marine environments. Laboratory studies seem appropriate here, as well as field studies of the distribution and stability of these chemicals in natural environments, particularly in interstitial waters of soils and of coastal marine sediments. It is in studies of these environments that geochemistry can contribute to knowledge of the pathways and sinks of pesticides and related chemicals.

SUMMARY

Among the many areas in which geochemists can contribute to solution of environmental problems are the following:

- Assessment of the long-term effects of man's interference in natural element cycles on a global scale. This should be regarded as a priority item. Steady-state and time-dependent models need to be developed for element cycles, particularly for those whose rates of addition to the earth's surface have been significantly increased by man's activities. Assessment of natural sources, transport paths, fluxes, sinks, and reaction rates of elements can provide baselines for evaluation of the effects of the increased rates of element addition to the environment from man's activities. These models can help governmental agencies to develop rational policies to control the national and global environment.

• Evaluation in natural environments of sources, transport paths, fluxes, sinks, and reaction rates of elements regarded as potential pollutants. Examples of such natural environments are coastal estuaries, soils, lakes, and rivers. Mechanisms of transport of pollutants from shallow to deep marine waters also need more investigation. These studies will be particularly useful in establishing regional, as opposed to global, policies. Inorganic and organic geochemists and specialists from other fields, such as ecology, soil science, and marine chemistry, should find these studies to be fertile areas of interaction.

NOTE:

As indicated in the summary of this report, the relations of geochemistry to environmental concerns have been covered only in the broadest terms. An additional report would be required to indicate the scope of the present and probable future involvement of geochemists in environmental studies. A few representative projects that are at present using the skills and background knowledge of geochemists may help to suggest the contents of such a report. They are as follows:

• Geochemistry as related to health and disease.
• Joint studies with biologists on local effects of sewage outfalls.
• Use of isotope ratios to trace sources of pollutants.
• Studies of immobilization of radioactive elements in minerals for safe underground storage.
• Local studies of increase of weathering rates from atmospheric sulfuric acid.
• Studies of diffusion of pollutants in and out of sediments.

SELECTED READINGS IN GEOCHEMISTRY AND ENVIRONMENTAL CONCERNS

Dryssen, D., and D. Jagner (eds.). The changing chemistry of the oceans. New York: Wiley-Interscience, 1972. 365 p.

Hood, D. W. (ed.). Impingement of man on the oceans. New York: Wiley-Interscience, 1971. 738 p.

Hopps, Howard C., and Helen L. Cannon (eds.). Geochemical environment in relation to health and disease. Ann. N.Y. Acad. Sci. 199, 1972. 352 p.

Massachusetts Institute of Technology. Inadvertent climate modification: Report of the study of man's impact on climate. Cambridge: M.I.T. Press, 1971. 308 p.

Massachusetts Institute of Technology. Man's impact on the global environment: Report of a study of critical environmental problems. Cambridge: M.I.T. Press, 1970. 319 p.

Ocean Science Committee. Marine environmental quality: Suggested research programs for understanding man's effect on the oceans. Washington: National Academy of Sciences, 1971. 107 p.

Geochemistry
and Natural
Resources

It is obvious that our civilization is based on the availability of cheap raw materials, that these are being used at an ever increasing rate, and that the exhaustion of at least some important natural resources is in sight. It is, perhaps, less obvious that raw materials are natural chemicals, that their study falls in part within the field of geochemistry, and that geochemists continue to make significant contributions both to our understanding of the origin of natural resources and toward the more efficient search for the resources of the future.

Natural resources are frequently grouped into metals, nonmetals, and energy resources. Like most classifications, this is somewhat arbitrary and misleading, but generally useful. The metals include the three giants—iron, aluminum, and copper—and some 35 others whose rate of use is considerably less. Uranium is one of the relatively rare metals, but as a major energy source its importance is out of all proportion to its abundance. The nonmetals include limestone, gypsum, salt, sulfur and many other elements and compounds. Traditionally the energy resources have consisted largely of the fossil fuels—coal, oil, and natural gas. Oil shale and tar sands are easily fitted into this category, but uranium belongs more realistically among the metals, and water among the nonmetals.

Our response to the increased need for natural resources has been predictable and reasonable. An increased need for a resource is followed almost immediately by an increased price and rate of exploration. The need for uranium after World War II led to the discovery of extensive new uranium reserves. The intensive exploration of the continental shelves for oil and gas is yielding excellent results, but may represent the last major frontier in the search for these major fuels.

Most natural resources are nonrenewable in the sense that a deposit can be mined only once. Metals can, however, be recycled, and this will surely be done more extensively in the future. Complete recycling is not economically feasible, however, and the increased demand for new resources will exert great pressure for the development of less expensive mining and extraction methods, for the substitution of more abundant for less abundant resources, and for increasing the number of elements and compounds extracted from mined material.

GEOCHEMICAL EXPLORATION

The role of geochemistry in the field of natural resources is apt to be twofold: Basic research on the origin of ore deposits will contribute to our understanding of these rather unusual rocks; applied research will tend to use these insights in mineral exploration and mining technology. The recognition of bertrandite [$Be_4 Si_2 O_7 (OH)_2$] in tuffaceous (volcanic ash) rocks of the Thomas Range in Utah as a new type of beryllium ore, and the development of more effective mining methods, such as in-place leaching and hydrothermal mining, are examples of geochemical research successfully applied. Geochemical exploration has played an increasingly important part in the discovery of metallic ore deposits and has become a standard technique in the repertoire of exploration tools used by the mining industry. So far, most geochemical exploration has been based on the observation that abnormally high concentrations of the ore metals and sulfur dioxide are usually present in rocks well beyond the limits of commercial ore bodies, and that the weathering of ore bodies disperses the ore metals in plants and soils, and thence in the nearby atmosphere and in the water and sediments of rivers draining mineralized areas. Geochemical prospecting is frequently used to distinguish geophysical anomalies related to ore bodies from those above unmineralized or poorly mineralized ground. The theoretical framework for such prospecting has been relatively simple, and perhaps more effective and somewhat more subtle techniques will follow from a better understanding of the origin of mineral deposits.

GEOCHEMISTRY OF ORE DEPOSITS

The five most prevalent kinds of mineral deposits are magmatic, hydrothermal, weathering residues, sedimentary, and oceanic.

Magmatic Ore Deposits

During the crystallization of magmas (rock melts) produced near the base of the crust or in the upper mantle, economically important quantities of chromite and compounds of platinum and platinum-group metals may settle out gravitationally near the base of the intrusive mass. This is the origin of chromite in the famous Bushveld Complex of South Africa, which contains hundreds of millions of tons of chromite ore and enough platinum to ensure the world's supply for centuries.

Some magmas contain enough dissolved sulfur to permit sulfide melts rich in iron, copper, and nickel to separate during crystallization. The chemistry of this process and the complex processes which accompany the cooling of such sulfide melts are finally being clarified, and the results have been used to understand the development of major nickel–copper deposits such as those near Sudbury, Canada.

Hydrothermal Ore Deposits

Many magmas contain several percent water. Some of this water is retained in the crystallization of hydrous silicate minerals. However, most of the water boils off. If the magma contains at least several tenths of a percent of chloride, the solutions which boil off are highly saline and are capable of extracting metals such as lead, zinc, and manganese almost quantitatively from the magma. Sulfur is almost always present in magmas, and escapes in part with aqueous solutions. On cooling, some of the metals (iron, tin, and tungsten, for instance) are precipitated as oxides, many others (such as copper, lead, zinc, silver, arsenic, and antimony) as sulfides, and a few (such as silver and gold) as the elements themselves. Extensive studies of the solubility of the ore minerals, of their stability fields, and of the reaction paths of solutions during cooling have finally led to an understanding of the mineralogy of ore deposits. Detailed radiometric dating has shown that the time interval between the emplacement of a magma and the formation of associated ore deposits is usually less than 10^5 years. However, measurements of the isotopic composition of oxygen and hydrogen in minerals and in samples of the ore-forming fluid trapped in ore and gangue minerals have shown that shallow groundwater can penetrate into hot intrusives and can participate in the process of metal extraction and ore deposition. In fact, the whole complex of questions surrounding the hydrology of such plumbing systems is still unsettled; the answers will have obvious implications for prospecting for this type of ore deposit.

In oceanic areas, seawater apparently intrudes into magmatic settings. The extensive stratiform copper, lead and zinc (Kuroko) deposits of Japan, the copper deposits of Cyprus, and the very extensive sediments rich in iron and manganese associated with mid-ocean ridges and rises are almost certainly related to the interaction of seawater with hot igneous rocks at or below the ocean floor.

In many ore deposits that are clearly deposited from hot aqueous solutions there is no, or at best a very tenuous, connection to magmatic processes. The large lead–zinc deposits of our own midcontinent area and of other geologically similar provinces seem to be related, at least in part, to brines circulating in large sedimentary basins. The extensive uranium deposits of the Colorado Plateau are also related distantly, if at all, to igneous activity. Again, continuing research in hydrology coupled with isotopic "fingerprinting" and the study of fluid inclusions and mineral solubilities may hold valuable clues to the origin of these deposits and for future mineral exploration.

Several attempts have been made to relate ore deposition to plate-tectonics and to the development of subduction zones. The general relationships are now apparent for a few types of deposits, but detailed models are still hampered by the forbidding distances between the near-surface environment of ore deposition and the subduction zone itself.

Residues of Weathering as Ore Deposits

During chemical weathering the alkali elements, the alkaline earth elements, and silica are normally removed preferentially from silicate rock. Residues are usually enriched in aluminum and iron. Extreme enrichment produces bauxites (rocks consisting largely of hydrated aluminum oxides), the most important aluminum ores. Very large quantities of high-grade bauxites exist, but these are clearly incapable of sustaining the aluminum industry far into the future. The industry will then probably turn to clays as a source of aluminum. The supply of these clays is so large that the locale of aluminum mining will be determined for a long time by the availability of cheap transportation and cheap power.

Flowing water frequently acts as an excellent sorter for the detritus of erosion and weathering. Gold, platinum, cassiterite (tin oxide), in addition to titanium and zirconium minerals, are sometimes so efficiently concentrated that river and beach sands and gravels can be mined profitably. Such placer gold was important in the opening of the western United States. The major gold fields of South Africa and the associated uranium mineralization are generally considered to be ancient placers that have suffered a certain amount of remobilization during later deep burial.

Sedimentary Ore Deposits

Most of the largest iron deposits were formed by precipitation from surface waters, probably marine. Most of the enormous iron ore deposits of the Lake Superior region, of Australia. South Africa. the U.S.S.R., and South America were formed in a surprisingly small interval of time about 2000 million years ago. Manganese is one of the other elements that are concentrated chemically in sediments, and manganese nodules with their high concentrations of nickel, cobalt, and copper have made the ocean floor of interest to prospectors. Phosphate, cycled up from the deeper parts of the oceans, has been deposited near areas of upwelling as a constituent of calcium phosphate, and subsequently has been sufficiently concentrated in many areas to warrant mining. Uranium is often present in these ores, but rarely in currently commercial quantities. The concentration of uranium in shales rich in organic matter is probably of greater economic importance. Rocks like the Chattanooga black shales of the eastern United States may well become enormous low-grade uranium ore deposits.

Ocean Water as Ore

The grade of most ores has been decreasing with time. Copper ores mined in the nineteenth century contained several percent of the metal. Today most copper ores contain considerably less than 1 percent copper, and it is likely that large open-pit deposits, containing only 0.3 percent copper will shortly be mined. For some elements the seemingly inexorable march toward lower grades will stop abruptly when the cost of extraction from solid rock equals that of extraction from ocean water. This has already happened to magnesium and bromine. The oceans represent a virtually inexhaustible supply of these elements, so that their future cost depends on extraction and marketing costs. The increasing need for fresh water in areas such as southern California will surely make desalination of seawater progressively more attractive. Large-scale desalination will produce salts of the elements dissolved in seawater as cheap by-products.

GEOCHEMISTRY OF ENERGY SOURCES

Fossil Fuels

During the twentieth century, the fossil fuels have been by far the most important energy sources. Oil and gas reserves are already seriously depleted so that, before long, coal is apt to assume its former position of honor. Oil shales and tar sands are on the verge of economic profitability and will surely be of major importance during the coming century. Current estimates of fossil-fuel exhaustion are uncertain, but it seems likely that fossil fuels will last much more than 300 years at the present projected rates of consumption.

Fossil fuels consist of organic matter that escaped destruction by oxidation. Today, more than 99 percent of the organic matter produced photosynthetically is lost by oxidative processes. The efficiency of oxidation has probably been very high for at least the last several hundred million years; coal, oil, and gas are a very small part of a small residue of the total organic productivity of the earth.

Coal consists largely of plant material that accumulated in swamps and has since lost a large fraction of its original combined hydrogen, oxygen, and nitrogen. The geometry of coal seams reflects that of the containing sediments, so that coal reserves may be estimated with some accuracy.

Conversely, oil and gas are products of the decomposition of organic matter in sediments that have moved sufficiently to make the identification of their source rocks difficult. World reserves of these fuels are thus hard to estimate with any degree of accuracy, but as they apparently account for less than 10 percent of the total fossil-fuel reserves, the large uncertainty in the reserve figures for oil and gas do not seriously affect estimates of total fossil-fuel reserves.

The sequences of chemical reactions leading to the production of oil and gas are roughly understood; organic geochemists are much concerned with these reactions, not only for their own sake but because they may lead to a more certain identification of petroleum source rocks and thence to the location of attractive new targets for oil exploration.

The abnormal concentration of certain trace elements in coal has been known since the pioneering work of V.M. Goldschmidt some 40 years ago. The environmental and resource implications of these concentrations have been taken seriously only during the past few years. Mercury released during the burning of coal apparently contributes a sizable fraction of the total input of this metal into the environment, and more attention should and will be paid to the use or at least the containment of this element and of several other trace substances in coal.

Geothermal Energy

Geothermal energy accounts for only a small fraction of the world's energy sources, but it is of importance in New Zealand and Italy and is being developed actively in Japan, Mexico, and in the western United States. The feasibility of extracting heat from surface fluids that are pumped underground and then returned to the surface is being debated; whatever the outcome, geothermal power is and will continue to be attractive because it is relatively nonpolluting.

Most geothermal fluids consist of rainwater that has penetrated into areas of abnormally high thermal gradients, has been heated, has reacted with rocks at depth along its migration path, and has returned to the surface as dilute solutions of chloride, bicarbonate, and sulfate salts. In some geothermal areas, such as the Salton Sea area of California, the fluids emerge as highly saline brines and may be of more use as sources of chemicals than of power.

The physics and chemistry of geothermal systems have been studied perhaps most intensively in New Zealand. The results of these investigations have obvious implications for finding new geothermal areas and for understanding and location of systems that give rise to hydrothermal ore deposits of a large number of metals.

SUMMARY

Geochemical research during the last two decades has contributed materially to our understanding of the origin and development of a variety of natural resources and to the development of useful exploration techniques. Perhaps the most significant discoveries have been made in the field of hydrothermal geochemistry by the combined application of solution chemistry at elevated temperatures and pressures, the study of the distribution of both stable and radiogenic isotopes, radiometric age determination, and the analysis of fluid inclusions in ore and nonore minerals. It seems likely that many of the present outstanding problems can be solved during the coming decades and that their solution will influence mineral exploration. The discovery of several new types of ore deposit during the past decade has encouraged thinking in mineral exploration along unconventional lines, and it is likely that a better understanding of ancient environments will have a similar effect.

Experimental Techniques and Facilities

The past decade has seen a sharp improvement in available experimental techniques. In many instances the advances appear to have reached plateaus sufficient to meet many future needs. Electron microprobes, for example, may now be fully automated so that a total of 10–12 elements may be analyzed on preselected spots in as little as several minutes per spot. Special features, such as automatic peak location and drift corrections, lead to improved accuracies, often of a few percent. One may now visualize undertaking projects that require hundreds of high-quality analyses and completing the analytical work in a few days.

Measurement of isotopic ratios by mass spectrometry has also experienced dramatic improvements in precision, speed, and automation. For example, strontium isotopic ratio analysis may be routinely made at the 0.01 percent precision level. Since this is in most instances lower than the "geologic noise," further improvements are unlikely to be usable (or necessary) for most terrestrial problems. Similarly, with oxygen isotopic measurements, geologic effects and interpretations of models generally impose larger uncertainties than the analytical precision. Age-dating studies are also in general not limited by analytical precision, but by geologic problems. For example, improved uranium/lead dating on zircon, which promises a 1-million-year age resolution for even old Precambrian rocks, may not realize this level of resolution because of a lack of understanding of the geologic effects on this dating system.

Development of new techniques in some areas is critically needed, and in other areas will occur naturally as fields outside geochemistry provide them. For example, experience has shown that as new instruments and techniques are developed in the fields of organic chemistry and in biochemistry, they are rapidly adopted for use in organic geochemistry. Gas–liquid chromatography was introduced to organic geochemical studies and was a major step toward understanding the complex mixtures of organic molecules found in geological materials. This technique suffered, however, because the detector was not specific for particular molecules. A partial solution was obtained by using a mass spectrometer for a detector, and such a combination is now used routinely, identifying, for example, 50 or more components in a mixture. The limitation of this approach was found to be in handling the vast amount of spectral data obtained. Here the use of automatic data-handling systems becomes important. Systems for comparison of unknown spectra with large banks of reference spectra have been put into operation, and the automatic identification of individual components of highly complex mixtures is rapidly becoming a reality. Proliferation of data under such circumstances may become a burden, and the ultimate limitation will probably lie with the ability of the individual scientist to assimilate and interpret data.

Nuclear magnetic resonance spectrometry is another technique that is used routinely and indispensably in organic chemistry, and improvements in sensitivity may allow major applications in organic geochemistry. It will then become possible to detect subtle details in the structures of fossil steroids and their transformation products, thus providing new information on paleobiochemistry and reaction pathways.

Another instrument, designed in a field outside geochemistry, but which may well undergo major development by geochemists, is the ion probe (ion microprobe mass spectrometer). The electron microprobe has revolutionized the fields of phase equilibria and petrology during the past decade, and the field of trace element geochemistry is critically in need of a similar revolution. The ion microprobe, which allows concentrations and isotopic ratios to be determined on most elements (including the low-mass elements) at the parts-per-million concentration level and over microareas in the micron range, has already shown great potential in studying geochemical problems. Any real impact, however, must await the placing of a number of these machines in geochemical laboratories and the solution of a number of technical problems. The instruments are relatively expensive, costing around $300,000, and some consideration of centralized facilities may be required.

Although ion-probe techniques will provide much of the urgently needed microanalysis capability for trace element partition studies, alteration studies, and diffusion and reaction-rate studies, theoretical understanding of trace element partitioning will still be hampered by the inability to determine specific structural locations of the trace elements in their hosts. To understand trace element partitioning properly, it is important to know the sites in which the trace elements are being substituted. One promising technique for such studies may be Auger spectroscopy which, in utilizing the decay of excited outer-shell electrons, is sensitive to trace elements and to slight variations in the structural positions of these elements. This technique is unfortunately only able to "see" the outer few angstroms of a crystal, so the material surfaces must be freshly prepared and undisturbed.

Finally, an area of geochemistry seriously held back by technique limitations and for which the prospects of necessary developments is unclear, is that of high pressure–high temperature phase equilibria. Although studies of phase equilibria at conditions relevant to the upper mantle and crust are now fairly routine, study of the rest of the earth is beyond present capabilities. Shock wave studies are of considerable use, but do not allow the detail necessary to delineate the many pressure–temperature composition facies possible for the mantle and core. The development of equipment and techniques for this major area of study will be a formidable task and will probably require investment and support at the national level for a centralized facility.

Data and
Sample
Accessibility

A recurring problem in geochemistry, as in other expanding fields of science, is maintenance of the availability of good data (including information on their validity) and analyzed samples. A detailed status report or list of recommendations is beyond the capability (and charge) of this Panel. Nevertheless, we feel that the following recommendations and comments warrant detailed study as soon as possible.

DATA

1. A sizable number of data collections already exist. A summary list of these collections should be prepared covering the following points:

a. The types of data assembled.

b. The criteria used to accept or reject data, and the way in which information on data quality (precision and accuracy) are incorporated in the collections.

c. The use already made of each collection and its assemblers' view of its accessibility and the effectiveness of its utilization.

2. Partly on the basis of the results of (1), a decision should be

made as to whether centralized data banks are warranted and to what extent. If they are warranted, consideration must be given to the following:

a. The organization best equipped to handle the collections.

b. The contents, formats, and criteria for data acceptance of these data banks.

3. In the absence of centralized data banks, or if they are found to be undesirable, the accessibility of published data should be improved by better documentation of samples. The following points may be helpful:

a. Journals should not accept data for publication unless the samples are adequately documented. At the very least, the location and geologic setting, gross description, present location, and specific identifying number or code for each sample should be included.

b. The use of United States National Museum (USNM)-assigned identifying numbers, similar to those used by micropaleontologists, has much to commend it to geochemists. Under such a system, a researcher would obtain a USNM number for each sample before publication. This number would then be published with the analysis, and the sample (or a split of it) would be deposited with the USNM when the work was completed. In this way there would be more likelihood of supplementary work, which greatly increases the value of analyzed material.

SAMPLES

1. The availability and effective use of samples is closely related to the inclusion of adequate documentation with published analyses, as discussed in the previous section. If published analyses are inadequately identified, no curatorial system can make the samples really accessible.

2. The present status of analysed samples should be documented. The following points need to be considered:

a. Location of samples.

b. Curatorial system.

c. Procedure to be followed by someone seeking a split of an analysed sample (basically, with whom to get in touch).

d. Identification and general description of major collections with comments on their accessibility.

3. Many of the classic collections of analyzed samples, on which

major geochemical syntheses and conclusions are based, are maintained only by their collector. Thus, there is a serious possibility that such collections will be lost or at least become inaccessible on the retirement or death of their mentor. As an example, the Poldervaart collection of rocks at Columbia University has already fallen into disarray in the 3 years since Poldervaart's death. The loss of such collections is sufficiently damaging to the field of geochemistry to make it necessary to develop some mechanism to assure their preservation and long-term accessibility.

4. As previously stated, serious consideration should be given to depositing splits of all analyzed samples with a central repository, perhaps the USNM. Even if such a procedure is not acceptable to geochemists, insistence that each sample analyzed have a unique identifying number (a policy that journal editors could easily police) would greatly reduce the present confusion and encourage additional multiple analyses of key samples.

5. Funding agencies should be encouraged to view curatorial expenses as a basic research cost. They should also insist that analyzed material be accessible to subsequent workers.